LABORATORY MANUAL

Arthur N. Strahler
Alan H. Strahler

LAB MANUAL FOR PHYSICAL GEOGRAPHY

Alan H. Strahler
Boston University

Arthur N. Strahler
Columbia University (emeritus)

JOHN WILEY & SONS, INC.

Cover Photo: © Corbis Digital Stock

To order books or for customer service call 1-800-CALL-WILEY (225-5945).

ISBN 0-471-47670-6

Printed in the United States of America

10 9 8 7 6 5 4 3 2 1

Printed and bound by Courier Kendallville, Inc.

Table of Contents

*Chapter titles are those of the Strahler & Strahler textbook, INTRODUCING PHYSICAL GEOGRAPHY, THIRD EDITION, 2004, John Wiley & Sons, Inc. New York.

Chapter 5 Winds and Global Circulation

Chapter 6 Weather Systems

Chapter 7 Global Climates

Chapter 8 Biogeographic Processes

Chapter 9 Global Biogeography

INTRODUCTION

This *Laboratory Manual for Introducing Physical Geography, Third Edition,* builds on your knowledge of physical geography developed through class lectures and study of your textbook. Not only will you put this knowledge to use in solving problems based on course topics, but also in reaching out in new directions to explore special topics not covered in the text. Some of these exercises are geographical journeys to study exemplary environmental features—such as landforms, climates, forest types, or soils—in various parts of the globe. Other exercises are journeys of the mind into methods of doing scientific research; they require you to analyze sets of data in search of a better understanding of how the processes of nature act in varied global environments. In this way you will be taking part in a sampling of choice topics from several of the many fields of physical geography that are currently under intensive scientific investigation.

Each exercise is keyed to specific pages, illustrations, maps, or tables in your textbook. Although many questions contain all of the data needed to solve the problems and answer the questions, for some of them you will need to refer to the textbook for working data. In some cases, special explanations are provided in the manual for particular topics that extend beyond what your textbook covers.

Your exercise manual provides ample space in which to write out answers in full. Blank graphs for plotting data are provided, as well as diagrams and maps on which information is to be inserted. The perforated and punched pages can be easily removed for convenience in doing the work and submitting it for evaluation.

You will need to have on hand a few essential tools and materials. Drawing equipment includes the following: Dividers with sharp points for measuring and laying off distances on maps and graphs. (A well-sharpened pencil compass will do.) Triangle of clear plastic, preferably 30°/60°/90°, at least six inches on the longest side. Protractor, spanning 180°. Color pens and/or pencils (red, blue, green, orange), needed for plotting point and line data and for coloring areas on maps and graphs. A dozen or so sheets of tracing paper ($8\frac{1}{2} \times 11$ in.), required for several exercises.

A student atlas should be at hand as you work, and is essential for some of the exercises. (Recommended: a recent edition of *Goode's World Atlas*, Rand McNally, Publishers.) In addition, a good globe showing both physical and political features will be very helpful.

NAME _____   DATE _____

Exercise 2-A Insolation and Latitude

[Text p. 57, Figure 2.6.]

In this exercise we investigate the way in which incoming solar energy, or insolation, diminishes from full strength at the equator to zero at the two poles. The conditions specified are that the earth is at either the vernal equinox or the autumnal equinox, when the sun's noon rays strike the equator at an angle of 90° (a right angle) with respect to a horizontal flat surface. We assume that no atmosphere exists to diminish the intensity of insolation. Examine Figure 1.18 in your textbook to get the picture.

The way in which insolation varies with angle made by the sun's rays is illustrated in Figure 2.6 of your textbook. Study the figure closely and read the text that goes with it (p. 57).

Figure A develops in great depth the relationship between latitude and the intensity of insolation. Notice, first, that the angle of latitude plus the angle between the sun's rays and the horizontal surface ("angle of incidence") always equals 90°:

Angle of latitude	Angle of incidence	Right angle
0°	90°	90°
30°	60°	90°
60°	30°	90°
90°	0°	90°

Figure A

Intensity is given as percentage of the maximum possible, which is 100%, and applies at the equator. At either pole, the sun's rays barely graze the surface, but do not strike it, so there the value is 0%. In Figure A, right triangles are drawn at 30°N and 60°N. Side AB cuts across the sun's rays at right angles and represents 100% intensity. The hypotenuse of the triangle, side AC, represents the horizontal ground surface. As latitude increases, AC becomes longer, indicating that the insolation is becoming spread out over a larger area, i.e., that it is becoming less intense. Notice that in these triangles, the angle A is the same as the latitude angle; the angle C is the same as the angle of incidence.

The method we use in this exercise is to construct triangles like those in Figure A for the latitudes shown in Figure B, and on them make a direct measurement of the length of each hypotenuse (AC). Use Diagram C for this purpose. You will need a protractor to construct the latitude angles correctly. Diagram C provides a scale with which to measure length of each hypotenuse. One unit on this scale is equal to the length of AC, which is constant for all latitudes. Enter these measurements in the table provided below and calculate the percentage. The problem is solved for latitude 45°.

Lat.	AB	AC	AB/AC	×100
0°	1.00	1.00	1.00	100%
15°	1.00	——	——	——
30°	1.00	——	——	——
45°	1.00	1.41	0.71	71%
60°	1.00	——	——	——
75°	1.00	——	——	——
90°	1.00	——	——	——

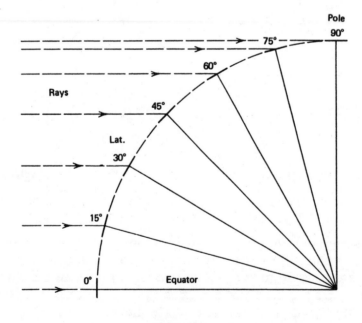

Figure B

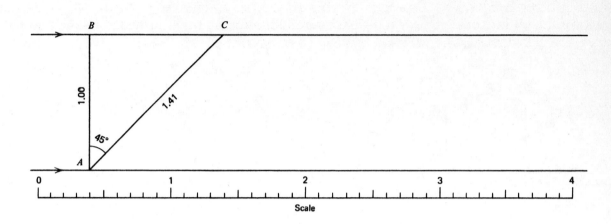

Figure C

Optional method using trigonometry. (Special credit.)

The values for each latitude can be found more easily and accurately by use of trigonometry, as shown in Figure D. Using a table of the values of the cosine of an angle, fill in the missing values in the table below.

Lat.	Cos lat.	×100
0°	1.000	100.0%
15°	_____	_____
30°	_____	_____
45°	0.707	70.7%
60°	_____	_____
75°	_____	_____
90°	_____	_____

By trigonometry:
At lat. 45° N

$$\frac{AB}{AC} = \cos 45° = 0.707$$

Figure D

NAME _____ DATE _____

Exercise 2-B The Annual Cycle of Insolation

[Text, p. 57–61, Figures 2.7, 2.8.]

The annual cycle of insolation at a given location on the globe depends on two factors: (a) the latitude at which the observer is located and (b) the sun's changing angle above the horizon at noon.

You are asked to show the sun's changing angle throughout the year on Graph A on the following page. The angle used on the vertical scale of the graph is known as the *sun's declination*, which for our purposes can be thought of as that global parallel of latitude over which the sun at noon is in the zenith position (straight up in the sky). This condition occurs at the equator twice yearly on the dates of the two equinoxes (March 21 and September 23). The sun is directly overhead at noon over the tropic of cancer, lat. $23\frac{1}{2}°$ N., at summer solstice, June 21; over the tropic of capricorn, lat. $23\frac{1}{2}°$ S, at winter solstice, December 22.

Table A gives the sun's declination angle at ten-day intervals throughout the year. Plot these points on the blank graph (A) and connect them with a smooth curve. This is a kind of mathematical curve called a *sine curve* or *sine wave*. Label the horizontal line of 0° as "Equator." Draw horizontal lines at $23\frac{1}{2}°$ N and $23\frac{1}{2}°$ S; label them "Tropic of Cancer" and "Tropic of Capricorn."

Table A Sun's Declination Throughout the Year

Date		Declination (degrees)		Date		Declination (degrees)	
Jan	1	23	S	Jul	10	$22\frac{1}{2}$	
	10	22			20	21	
	20	20			30	$18\frac{1}{2}$	
	30	$17\frac{1}{2}$		Aug	10	16	
Feb	10	15			20	$12\frac{1}{2}$	
	20	11			30	9	
Mar	1	8		Sep	10	5	
	10	$4\frac{1}{2}$			20	$1\frac{1}{2}$	
	20	$\frac{1}{2}$			30	$2\frac{1}{2}$	S
	30	$3\frac{1}{2}$	N	Oct	10	$6\frac{1}{2}$	
Apr	10	$7\frac{1}{2}$			20	10	
	20	11			30	$13\frac{1}{2}$	
	30	$14\frac{1}{2}$		Nov	10	17	
May	10	17			20	$19\frac{1}{2}$	
	20	20			30	$21\frac{1}{2}$	
	30	22		Dec	10	$23\frac{1}{2}$	
Jun	10	23			20	$23\frac{1}{2}$	
	20	$23\frac{1}{2}$					
	30	23					

The intensity of insolation throughout the year at various latitudes is given in Table B. The value of one unit used in this table is 889 gram calories per square centimeter, or 889 langleys. (Note that in your textbook, the langley has been replaced by watts per square meter, but the curves you draw will take the same form in either set of units.) On the blank graph (B), plot these monthly values for the following latitudes: 0° (equator), 20° N, 40° S (south), 60° N, 90° N. Enter the data point in the middle of the month as shown on the partially drawn line for 60° N. Connect the points with a smooth curve and label the latitude.

Table B *Insolation Throughout the Year*

Latitude		Jan.	Feb.	Mar.	Apr.	May	Jun.	Jul.	Aug.	Sep.	Oct.	Nov.	Dec.	Total for year
	90	...	...	1.9	17.5	31.5	36.4	32.9	21.1	4.6	...	...	...	145.9
	80	...	0.1	5.0	17.5	30.5	35.8	32.4	20.9	7.4	0.6	...	...	150.2
°N	60	3.0	7.4	14.8	23.2	30.2	33.2	31.1	24.9	16.7	9.0	3.8	1.9	199.2
	40	12.5	17.0	23.1	28.6	32.4	33.8	32.8	29.4	24.3	18.4	13.4	11.1	276.8
	20	22.0	25.1	28.6	30.9	31.8	32.0	31.8	30.9	28.9	25.8	22.5	20.9	331.2
Equator		29.4	30.4	30.6	29.6	28.0	27.1	27.6	28.6	30.1	30.2	29.5	28.9	350.0
	20	33.8	32.2	29.0	24.9	21.2	19.6	20.5	23.7	27.7	31.1	33.3	34.1	331.1
	40	34.8	30.4	23.9	17.4	12.5	10.4	11.6	15.8	21.9	28.5	33.6	36.0	276.8
°S	60	33.0	25.3	16.0	8.1	3.3	1.7	2.7	6.5	13.6	22.6	31.1	35.3	199.2
	80	34.2	20.5	6.3	0.3	...	...	...	...	3.8	16.0	31.0	38.1	150.2
	90	34.7	20.7	3.2	...	...	...	...	...	1.0	15.6	31.5	38.7	145.4

(1) Study the insolation curves in relation to the annual curve of the sun's declination. Compare times of maximum and minimum values at 60° N and 40° S. Explain how these two curves differ in timing.

_____

_____

_____

_____

_____

_____

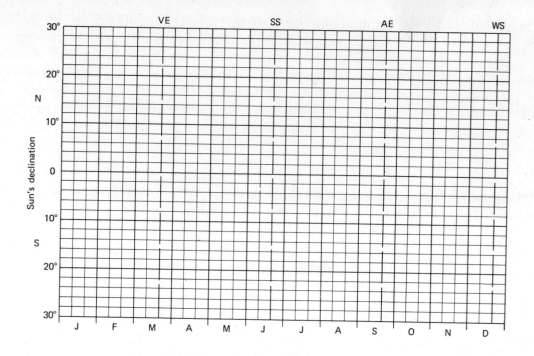

Graph A

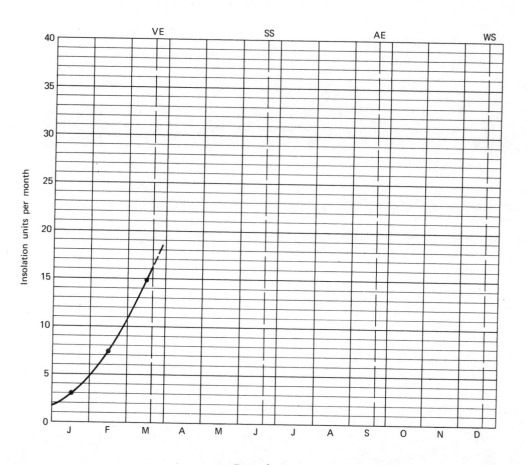

Graph B

(2) Explain how the maxima and minima of insolation at the equator are related to the curve of the sun's declination.

_____

_____

_____

(3) What disadvantage lies in the use of calendar months for analysis of the yearly cycle of insolation? Cite specific figures from the data table.

_____

_____

_____

_____

NAME _____   DATE _____

Exercise 2-C Duration of Sunlight at Different Latitudes

[Text p. 57–61, Figure 2.7.]

How long is the sun above the horizon at a given latitude and season? Is there romance in the "Land of the Midnight Sun"? Tourists by the thousands take the North Cape cruise to the northernmost tip of land in Norway, 71°N. There, if the weather cooperates, you can take a picture of the sun well above the horizon at midnight. The duration of daylight is an important factor in many aspects of everyday life, especially at high latitudes—the arctic and polar zones—where summer-to-winter inequalities of day length and night length are very great.

(1) Figure 2.7 of your textbook shows the path of the sun in the sky at 40°N latitude, which is about the latitude of Philadelphia, Indianapolis, Denver, and Salt Lake City. By simple addition, using the given times of sunrise and sunset, or by counting the hour intervals along each path, give the approximate number of hours of sunlight daily for each of the following points in the yearly cycle:

	Month and day	Number of hours
Equinox	March 21, Sept. 23	____
Summer solstice	June 21	____
Winter solstice	Dec. 21, 22	____

Figure A, a graph on the next page, shows the number of hours each day that the sun is above the horizon for selected north latitudes from the equator to the arctic circle. Check the number of hours you entered in the above blanks for 40°N by reading the hours from the graph for that latitude.

(2) On the graph, sketch in the curves for latitudes 70°, 80°, and 90°N.

(3) At the north pole, for how many consecutive months (or days) will the sun be above the horizon continuously for 24 hours of the day?

_____ months; _____ days

Same for the duration of darkness (defined as sun below horizon).

_____ months; _____ days

(Note: Twilight, which supplies important amounts of illumination when the sun is within a few degrees below the horizon, is not considered in making this answer.)

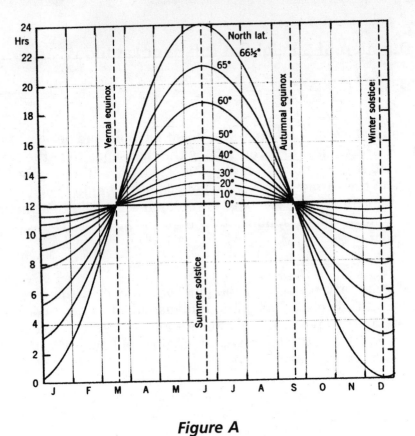

Figure A

(4) If we define "day" as the period when the sun is continuously above the horizon and "night" as the period when it is below the horizon, how many days and nights per year are experienced at the north pole?

_____ days, _____ nights

(5) What effect does length of daily sunlight period have on the summer growing season for food crops, such as grains and vegetables?

_____

_____

_____

_____

_____

(6) What factors of climate would tend to modify the effects of increase in number of sunlight hours with increasing latitude?

_____

_____

_____

_____

_____

_____

NAME _____ DATE _____

Exercise 5-A Converting Barometric Pressures

[Text p. 152–154, Figure 5.2.]

Barometric pressure is mentioned in most news media weather reports, if only to refer to "highs" and "lows." When the pressure value is stated, the unit is usually inches of mercury. The aneroid barometer, which is the kind most amateur weather buffs use, usually features the inch units, but some have a millibar scale as well. If you tune in on the National Weather Service continuous weather information broadcast on the VHF band, as do farmers, aircraft pilots, and yachters, you will hear pressure reported in millibars. Pressure in millimeters (or centimeters) of mercury is rarely used in weather information, but you're bound to encounter that scale when your physician reports your blood pressure as "130 over 70."

(1) As review, from text p. 154 and Figure 5.2, the value of standard sea-level

barometric pressure is: <u>29.92</u> in.; <u>760</u> mm; <u>1013.2</u> mb

(2) You are asked to make conversions of pressure readings from one scale to another, filling in the blanks of the table below. You should calculate the conversions using the formulas shown. The graphic conversion scale on the following page can serve only as a rough check on the inches/millibars conversions.

Inches	Millimeters	Millibars
30.12	___	___
___	710	___
___	___	1006
29.60	___	___
___	758	___
___	___	500

Conversion formulas:

1.0 in. (mercury) = 33.87 mb = 25.40 mm (mercury)

1.0 mb = 0.0295 in. = 0.75 mm

1.0 mm = 0.03937 in. = 1.3333.. mb

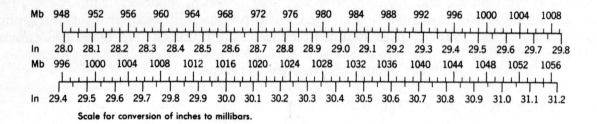

Scale for conversion of inches to millibars.

NAME _____   DATE _____

Exercise 5-B Pressure Versus Altitude

[Text p. 154, Figure 5.4.]

The curve of decreasing atmospheric pressure with increasing altitude, shown in text Figure 5.4, represents the average condition of the atmosphere. In technical language, it is the *U.S. Standard Atmosphere*. At any given time and place, the actual pressure curve would differ somewhat from the standard, and it is those small differences that are most important in weather phenomena, such as the interpretation and forecasting of cyclonic storms and weather fronts.

The table below gives data for the standard atmosphere in units of millibars and meters.

Pressure mb	Altitude m	Pressure mb	Altitude m
400	7,425	800	1,889
500	5,643	900	984
600	4,186	1000	106
700	2,955	1013	0

(1) Plot the data from the table on the blank graph provided. Connect the points as a smooth curve.

(2) Using the graph, estimate the standard pressure in mb for each of the following places. First, convert altitude from feet into meters using the values 1 foot = 0.305 m.

Canton, OH	1,030 ft	_____ m	_____ mb
Las Vegas, NV	2,030 ft	_____ m	_____ mb
Cheyenne, WY	6,100 ft	_____ m	_____ mb
Mt. Hood, OR	11,235 ft	_____ m	_____ mb
Mt. Whitney, CA	14,494 ft	_____ m	_____ mb

(3) Using the curve on the graph, estimate the amount of decrease in pressure in mb for 1 km of altitude increase at the following levels. Then calculate the percentage of decrease.

Between 1 km and 2 km:	_____ mb/km	_____ %
Between 2 km and 3 km:	_____ mb/km	_____ %
Between 6 km and 7 km:	_____ mb/km	_____ %

EX. 5-B

(4) Based on the percentages you have estimated, make a general statement about the rate of change of pressure with altitude:

_____

_____

_____

_____

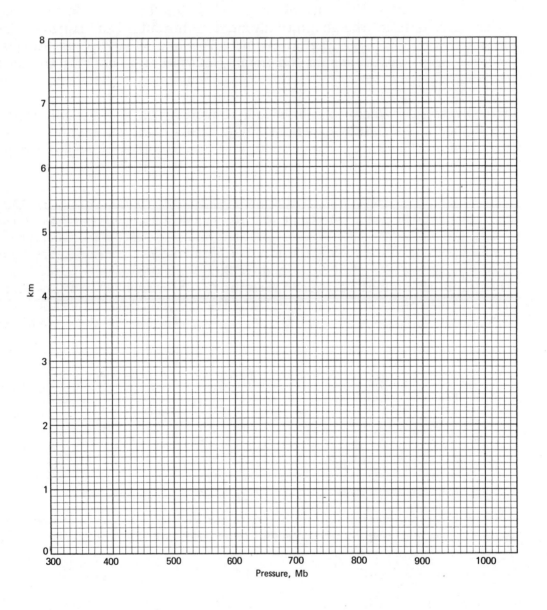

Exercise 5-C Isobars on the Surface Weather Map

[Text p. 154–155, Figure 5.7; p. 161–162, Figure 5.15.]

Ballot's Law How does the direction of the surface wind relate to barometric pressure? Many decades ago, when weather forecasting was in a primitive state, a Dutch meteorologist, Buys Ballot, formulated a rule-of-thumb (actually "rule-of-hand" would be more accurate) to keep things straight. He advised: "Stand with the wind at your back with your arms outstretched to either side, and the lower pressure will be toward your left." Called Ballot's Law, and since then almost forgotten, it works quite well, but with two provisions. First, you must be in the northern hemisphere; second, your left arm should be aimed midway between straight left and straight to the front. Try this out when your map exercise is done by imagining yourself standing on a wind arrow and looking in the direction in which the arrow is pointing. How would you amend Ballot's Law to apply in the southern hemisphere?

Drawing Isobars on the Weather Map Practice in drawing isobars on a weather map and figuring out how the surface winds move diagonally across them will increase your ability to read and interpret not only daily weather maps, such as those shown in Chapter 6, Figure 6.10, but also the global maps of pressure and winds, Figure 5.17.

The weather map on the reverse side of this page shows barometric pressures observed simultaneously at many National Weather Service stations. Pressures are in millibars, but only the last two digits are given. Thus, "16" designates 1016 mb; "96" designates 996. The station is located at the dot beside the number. Draw isobars for the entire map, using an interval of 4 mb. Isobars should run thus: 992, 996, 1000, 1004, 1008, 1012, etc. Label the isobars. Label lows and highs. In drawing the isobars, use a soft pencil lightly at first, allowing for many corrections. Then draw the final isobars as smooth, flowing curves. Finally, draw many short straight arrows across the isobars to show the directions of surface winds.

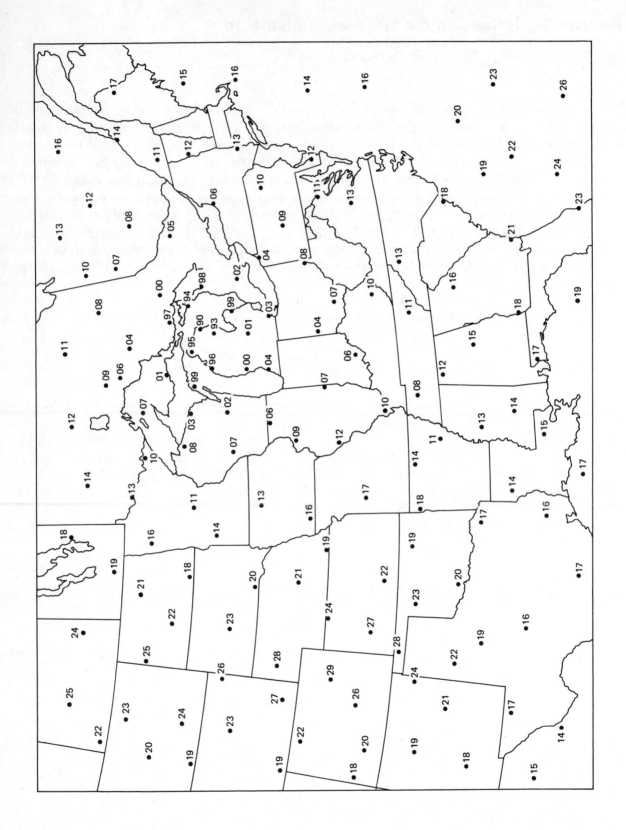

Exercise 5-D The Wind Rose and Global Wind Belts

[Text p. 156, 163–167, Figures 5.6, 5.16, 5.17, 5.18.]

The Beaufort Scale of Winds For this exercise we go back to the era of the great sailing ships—the clippers—that made use of the belts of prevailing winds to make their great transoceanic and circum-global cruises. Long before anemometers were available to measure wind strengths at sea, Admiral Sir Francis Beaufort of the British Navy in 1806 devised a system to standardize the estimation of wind force through its effects. It became known as the *Beaufort Scale of Winds*. It is reproduced on a following page. The *knot*, equal to one nautical mile per hour, is used in the Beaufort scale for marine applications. (One knot is equal to 1.15 miles per hour.) Under this system, wind speeds were given by numbers designating forces from 0 to 12, 0 being calm and 12 being hurricane force (above 65 knots). For example, a force of 5 was termed a "Fresh Breeze" and included velocities of 17–21 knots; a force of 9 was a "Strong Gale" (41–47 knots). Although of practical value when applied to the rigging of sailing vessels, the Beaufort scale was not adequate for modern scientific studies and has been abandoned. It is, however, used extensively on U.S. Navy Oceanographic Office pilot charts still available for study in libraries.

Wind Roses Shown below, at the left, is a special kind of graph known as a *wind rose*. It shows by arrows the average wind strength for each of eight 45-degree sectors of the compass. Data may be the long-term average for the entire year, or for a single calendar month over many years of observation.

(1) Refer to the right-hand wind rose below. Using the scale under it, estimate the percentages of time during which the wind blows from each of the sectors of the rose. Measure shaft length from the center of the circle. Total the percentages and subtract from 100 in order to determine the percentage of light air and calm. Enter your data in the table on the following page.

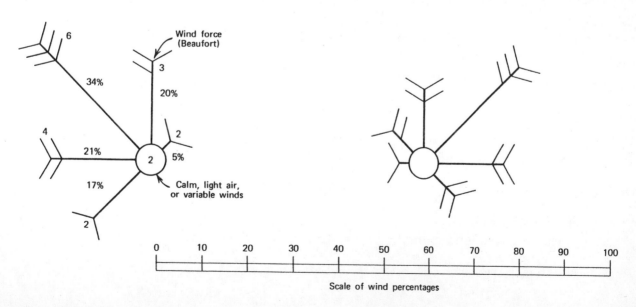

Scale of wind percentages

Sector	Percent of Time	Force
N	_____	_____
NE	_____	_____
E	_____	_____
SE	_____	_____
S	_____	_____
SW	_____	_____
W	_____	_____
NW	_____	_____

Total % : _____   % of calms: _____

Maps A and B are portions of two U.S. Navy pilot charts on which average wind strengths and directions for a particular month are shown for each five-degree square of latitude and longitude by means of the wind rose. Our exercise makes use of this graphic device to gain a closer understanding of the belts of prevailing winds.

(2) Referring to the two pilot charts, identify the wind belt from which the above data are taken. Estimate the latitude.

Name of belt: _____

Latitude: _____

(3) Examine the left-hand rose (previous page), on which percentages and wind force are labeled. From what wind belt is this rose taken? Estimate the latitude.

Name of belt: _____   Latitude: _____

(4) Identifying the major belts of prevailing winds. On the pilot charts, label the following belts of winds or calms. (Place labels vertically beside the right-hand margin of each map.) In bold lines, draw the boundaries between the belts. In the spaces below, enter the latitude range of each belt shown on these charts:

Doldrums	Lat. _____ to _____
Prevailing westerlies	Lat. _____ to _____
Southeast trades	Lat. _____ to _____
Subtropical belt of variable winds and calms	Lat. _____ to _____
Northeast trades	Lat. _____ to _____

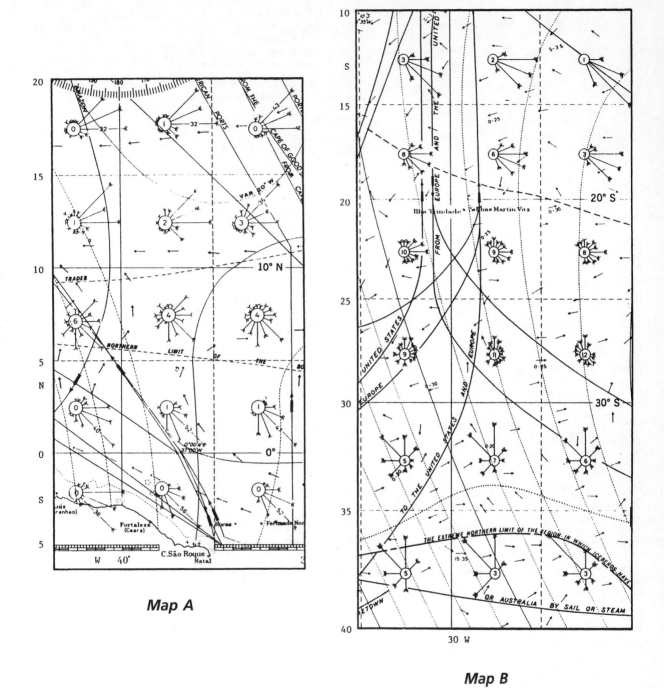

Map A

Map B

THE BEAUFORT SCALE OF WIND (NAUTICAL)

Beaufort No.	Name of Wind	Wind Speed		Description of Sea Surface	Sea Disturbance Number	Average Wave Height	
		knots	km/hr			ft	m
0	Calm	<1	<1	Sea like a mirror.	0	0	0
1	Light air	1–3	1–5	Ripples with appearance of scales are formed, without foam crests.	0	0	0
2	Light breeze	4–6	6–11	Small wavelets still short but more pronounced; crests have a glassy appearance but do not break.	1	0–1	0–0.3
3	Gentle breeze	7–10	12–19	Large wavelets; crests begin to break; foam of glassy appearance. Perhaps scattered white horses.	2	1–2	0.3–0.6
4	Moderate breeze	11–16	20–28	Small waves becoming longer; fairly frequent white horses.	3	2–4	0.6–1.2
5	Fresh breeze	17–21	29–38	Moderate waves taking a more pronounced long form; many white horses are formed; chance of some spray.	4	4–8	1.2–2.4
6	Strong breeze	22–27	39–49	Large waves begin to form; the white foam crests are more extensive everywhere. Probably some spray.	5	8–13	2.4–4
7	Moderate gale	28–33	50–61	Sea heaps up and white foam from breaking waves begins to be blown in streaks along the direction of the wind. Spindrift begins to be seen.	6	13–20	4–6
8	Fresh gale	34–40	62–74	Moderately high waves of greater length; edges of crests break into spindrift. The foam is blown in well-marked streaks along the direction of the wind.	6	13–20	4–6
9	Strong gale	41–47	75–88	High waves. Dense streaks of foam along the direction of the wind. Sea begins to roll. Spray affects visibility.	6	13–20	4–6
10	Whole gale	48–55	89–102	Very high waves with long overhanging crests. The resulting foam in great patches is blown in dense white streaks along the direction of the wind. On the whole the surface of the sea takes on a white appearance. The rolling of the sea becomes heavy. Visibility is affected.	7	20–30	6–9
11	Storm	56–65	103–117	Exceptionally high waves. Small- and medium-sized ships might be for a long time lost to view behind the waves. The sea is covered with long white patches of foam. Everywhere the edges of the wave crests are blown into foam. Visibility is affected.	8	30–45	9–14
12–17	Hurricane	above 65	above 117	The air is filled with foam and spray. Sea is completely filled with driving spray. Visibility very seriously affected.	9	over 45	over 14

Source: After R. C. H. Russell and D. H. Macmillan (1954), *Waves and Tides*, London, Hutchinson's Sci. and Tech. Publ., p. 54, Table 7; and N. Bowditch (1958), U.S. Navy Oceanographic Office Publ. No. 9.

NAME _____ DATE _____

Exercise 5-E Upper-Air Winds

[Text p. 167–172, Figures 5.23, 5.24, 5.25, 5.26.]

Knowledge of air flow at upper levels is crucial for weather forecasting. The patterns of lows and highs are relatively simple and can be shown in smooth circular or oval isobar patterns. Sharply defined changes in the direction of isobars are largely absent. The jet stream closely follows the pressure pattern, and it is the jet streams that "steer" the low-level cyclones.

Examine the isobaric map below. Depiction of upper-level pressure patterns is done somewhat differently than on the surface map. The imaginary flat base, or datum, of the map is a surface of equal standard pressure. What look like isobars are actually elevation contours. An example is the U.S. National Weather Service map reproduced below for conditions at 7:00 P.M., E.S.T., on a day in early April. The lines are contours drawn upon the 500-millibar pressure surface. In keeping with U.S. practice, contours are labeled in thousands of feet. You notice, however, that low elevation numbers correspond with low pressure of a conventional map; high numbers with high pressure.

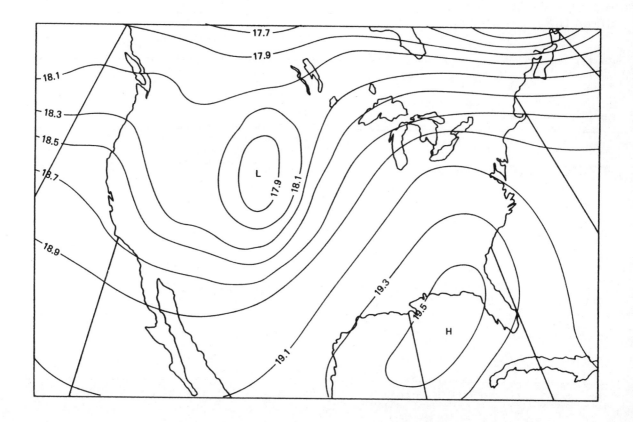

EX. 5-E

(1) Treating the contours as if they were isobars, draw numerous arrow points on the contours to show the direction of air motion. Before starting, examine Figure 5.25. Explain the relationship between wind direction and isobars at this level. Does Ballot's Law apply here?

_____

_____

_____

_____

_____

(2) Draw broad, sweeping arrows to show the probable position of the jet stream in two places on the map. (The jet stream would actually be at a much higher level of perhaps 35,000 ft.) Label a cyclone and an anticyclone.

(3) Imagine yourself piloting a private plane from San Francisco to New York at 18,500 ft altitude. You decide to try pressure-pattern flying, visibility conditions permitting, in order to keep the maximum tailwind in your favor. Describe your course and draw the path on the pressure map.

_____

_____

_____

_____

_____

_____

_____

NAME _____   DATE _____

Exercise 6-A Qualities of Various Air Masses

[Text p. 186–188, Figure 6.1, Table 6.1.]

It can be said with a high degree of truth that "the air mass makes the climate." An air mass affects the human senses simultaneously through both its temperature and its water vapor content. The climate classification system described in Chapter 7 is based in part on the kinds of air masses that are present in each season and how they determine whether there will be precipitation, and whether it comes as snow or rain. Know your air masses and your knowledge of weather and climate will be both profound and far-reaching in its applications.

Refer to text Table 6.1 for information on typical air temperatures and specific humidities for six major varieties of air masses. You are asked to plot these data on the graph provided here.

Compare this graph with Figure 4.4 on p. 123. On Graph A in this exercise we have copied a portion of that graph and extended the grid toward the left to include temperatures as low as −50°C. The curve on Graph A shows the maximum specific humidity (SH) on the vertical scale for any given air temperature (T) on the horizontal scale.

From text Table 6.1, read the values of T and SH given for each air mass and locate the appropriate point on the field of Graph A. Make a dot, surrounded by a tiny circle. Be as precise as you can in locating the point. Beside the point write in the symbol for the air mass: cA, cP, mP, etc.

(1) Recall that relative humidity, RH, expresses the ratio of the actual specific humidity of an air mass to the maximum specific humidity associated with the temperature of the air mass.

Which of the three low-latitude air masses (mE, mT, cT) probably has the highest relative humidity?

_____

Which has the lowest?   _____

How did you arrive at these answers? Explain.

_____

_____

_____

(2) Which three of the air masses would you expect to yield heavy showers or thunderstorms along a front? (Give symbols.)

_____

(3) In winter on Cape Code, Massachusetts, a powerful "nor'easter" (deep low off coast) brings ashore an air mass with T = 1°C and SH = 3 g/kg. Plot this point on the graph and label it **A**.

Assign an air mass symbol to this point: _____

Describe its position relative to the mP station you have already plotted, which is from the Pacific coast of the U.S. Northwest. Is this relationship to be expected? What does it mean?

_____

_____

_____

_____

(4) At 2 P.M. in midsummer in northern Ontario, Canada, a clear air mass registers T = 25°C and SH = 5 g/kg. Plot this point and label it **B**.

Assign an air mass symbol to this point: _____

Compare the position of this point to that of the winter cP airmass previously plotted. Explain your findings.

_____

_____

_____

_____

(5) In summer in Seattle, Washington, an air mass registers T = 15°C, SH = 8 g/kg. Plot this point and label it **C**.

Assign an air mass symbol to this point: _____

Compare this air mass with the two mP air masses previously plotted, both of them being winter air masses. Explain any differences you observe.

_____

_____

_____

_____

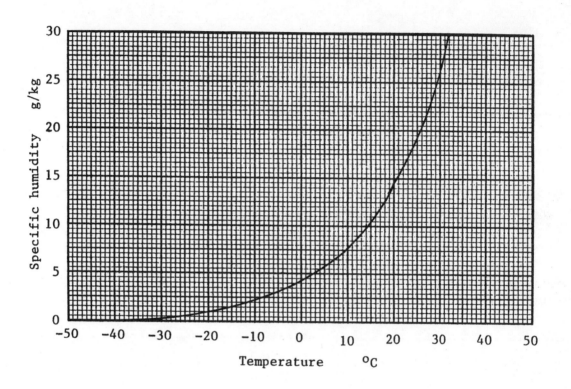

Graph A

NAME _____ DATE _____

Exercise 6-B World Weather on a Day in July

[Text, p. 194–196, Figures 6.11, 6.12.]

If you moved to Australia or New Zealand, your daily weather map might seem a bit crazy. Latitude gets higher as you move toward the bottom of the map, but the prevailing westerlies still move from left to right across the map and so do the cyclonic storms. The map, as compared with one in the northern hemisphere, is flipped top-to-bottom, but not left-to-right. This takes some getting used to, but will be easier after you have completed this exercise.

Reproduced below is text Figure 6.12, p. 195. This is a composite daily weather map for a typical day in July or August, when it is summer in the northern hemisphere and winter in the southern.

(1) How many low-pressure centers are shown on this map? _____

(2) How many high-pressure centers are shown? _____

(3) How many lows fall into each of the following classes?

      (a) Middle-latitude cyclones: _____

      (b) Tropical cyclones: _____

      (c) Weak lows of the equatorial trough: _____

(4) At what approximate latitude, or latitude belt, is each of the following located? (Give location of center of pressure cell.)

      (a) Equatorial trough (ITC):   _____

      (b) Subtropical cells of high pressure,   _____
             northern hemisphere:

      (c) Subtropical cells of high pressure,   _____
             southern hemisphere:

      (d) Middle-latitude cyclone centers,   _____
             northern hemisphere:

      (e) Middle-latitude cyclone centers,   _____
             southern hemisphere:

(5) Of the middle latitude cyclones shown on this map, how many are occluded, how many are open?

      Occluded: _____   Open: _____ (one indeterminate)

NAME _____ DATE _____

Exercise 6-C Air Masses Around the World on a Day in July

[Text p. 194–196, Figure 6.12.]

This exercise extends the study of the synthetic world weather map, Figure 6.12, to the identification of air masses in both hemispheres.

Place a sheet of tracing paper over the world weather map in Exercise 6-B. Trace the rectangular border of the map. Label each of the major highs and lows with the symbol for the air mass most likely to be associated with the pressure system. Use the standard air mass symbols, as in Exercise 6-A. Label the airmass found along the equatorial trough (ITC). Transfer the finished overlay to the rectangle below.

NAME _____   DATE _____

Exercise 7-A Seasonal Rainfall Contrasts in the Wet-Dry Tropical Climate

[Text, p. 235–236, Figure 7.14.]

The bar graph below shows seasonal rainfall contrasts for four stations in the wet-dry tropical climate (3). The rainfall of the wettest month is compared with the combined rainfall of the three driest months.

Station data are as follows:

Code	Station name	Location lat., long.	Mean annual total rainfall	Length of record, yrs.
BOM	Bombay, India	19°N 73°E	71.2 in. (181 cm)	60
DAR	Darwin, N.T., Australia	12°$\frac{1}{2}$S 132°E	58.7 in. (149 cm)	70
LUB	Lubumbashi, Zaire	12°S 27$\frac{1}{2}$°E	48.7 in. (124 cm)	17
PAR	Paraná, Brazil	12$\frac{1}{2}$°S 48°W	62.3 in. (158 cm)	19

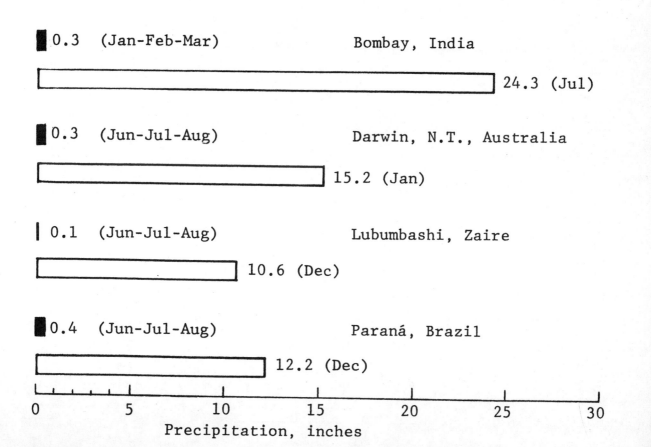

0.3 (Jan-Feb-Mar)   Bombay, India

24.3 (Jul)

0.3 (Jun-Jul-Aug)   Darwin, N.T., Australia

15.2 (Jan)

0.1 (Jun-Jul-Aug)   Lubumbashi, Zaire

10.6 (Dec)

0.4 (Jun-Jul-Aug)   Paraná, Brazil

12.2 (Dec)

0   5   10   15   20   25   30

Precipitation, inches

EX. 7-A

Locate each of these stations on (a) the world climate map, text Figure 7.2, and the world vegetation map, text Figure 9.3.

The port city of Bombay, long known to colonial Britons bound for military or civil service in India as the "Gateway to India," lies next to the Arabian Sea. Darwin, on the northern coast of Australia, was originally called Palmerston but was renamed in the honor of the great evolutionary biologist whose ship, *H.M.S. Beagle*, visited the spot in 1839. Lubumbashi was known during the colonial era as Elisabethville. A leading city in the Belgian Congo, it prospered as a smelting center for a booming copper-mining industry. Parana lies only about 200 mi (300 km) north of Brazil's spanking new capital city, Brasilia; they share much the same kind of climate, although Brasilia is on higher ground.

(1) Clearly, Bombay has the largest rainfall in the wettest month. Offer an explanation.

_____

_____

_____

_____

(2) Darwin's wettest month is January. Explain. Why is the June–August period so dry? (Hint: study the world pressure/winds maps of text Figure 5.17.)

_____

_____

_____

_____

(3) The maximum-month rainfalls of both Lubumbashi and Parana are low, compared with the other two. Give an explanation in terms of air masses.

_____

_____

_____

_____

NAME _____   DATE _____

Exercise 7-C The Subtropical Desert Temperature Regime

[Text p. 245–247, Figure 7.24.]

In this exercise we investigate the subtropical desert temperature regime. The vast subtropical desert of the southwestern United States attracts not only large numbers of recreation seekers, but also those retired persons looking for permanent residence in a land without a cold winter. For the latter group, the extremes of temperature are the most important climate factor. Air conditioning and space heating are an expensive necessity for much of the year. Water is a scarce commodity and must be imported from distant sources. Evaporation of soil moisture is intense in the hot months, disposing of much of that precious water with little benefit received.

Phoenix, the burgeoning capital city of Arizona, lies in the Sonoran Desert ($33\frac{1}{2}°$N, 112°W), close to the Gila River. Only a few miles away is Sun City, one of the largest and most popular retirement communities in the United States, its population approached the 50,000 mark in 1988. Would your parents like to spend their "golden years" in this environment? If they have such ideas, the temperature record for a typical July and January in Phoenix is well worth studying closely. Perhaps, instead, Sun City would be just "a nice place to visit."

Over a 52-year period of record, Phoenix had the following temperature statistics: Average daily maximum in July: 104°F. Average daily minimum in January: 39°. Highest recorded temperature: 118°. Lowest: 16°.

Graph A shows the daily high and low temperature for each month. These would be readings of the maximum-minimum thermometer in the standard shelter. The mean of the month and the average daily range are also shown.

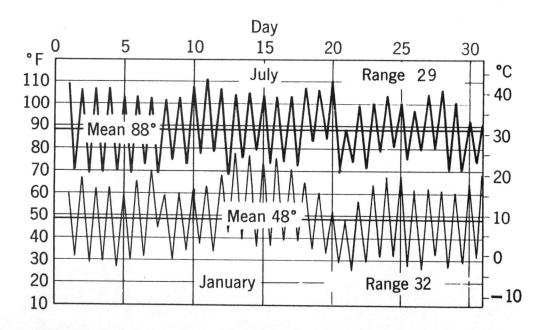

Graph A

Copyright © by Arthur N. Strahler

EX. 7-C

(1) Fill in the following data for July: (Use °F.)

Most consecutive days above 100°: _____

Highest temperature recorded: _____

Lowest temperature recorded: _____

Greatest range in one day: _____

(2) Fill in the following data for January:

Number of days with maximum 70° or over: _____

Number of nights below 35°: _____

Number of nights below 30°: _____

Greatest range in one day: _____

(3) Using the figures on the graph, about how great is the annual temperature range at Phoenix? How does this compare with the annual range at Yuma (text Figure 7.21) and at Wadi Halfa (Figure 7.17). Explain.

_____

_____

_____

_____

_____

_____

(4) For retirees owning a house in Sun City, might energy costs be a financial burden? Explain.

_____

_____

_____

_____

_____

(5) In the environment of Phoenix, how would temperature restrict summer outdoor physical activity, such as golf, tennis, or running?

_____

_____

_____

_____

(6) When air temperatures are over 100°, what level of relative humidity might be expected?

_____

_____

_____

NAME _____   DATE _____

Exercise 7-D Climographs of Midlatitude and High-Latitude Climates

[Text p. 246–262, Figures 7.20, 7.24, 7.27, 7.29, 7.32, 7.33, 7.36, 7.38, 7.41.]

This exercise is a continuation of Exercise 7-B and the same instructions apply. Using the latitude and longitude given, find the location of each station on the world climate map, Figure 7.2.

Plotting the data: As in your textbook climographs, monthly temperatures (°F) are plotted as points centered in each monthly column and connected by straight-line segments. Mean monthly precipitation amounts (inches) are plotted as bars, forming a step-graph.

Beside each climograph is a list of quantities to be entered in the blanks provided. The *mean annual temperature* (sum of monthly means divided by 12) is given with the station data. Calculate and enter the following statistics: *annual temperature range* (difference between highest and lowest monthly means); *annual total precipitation* (sum of the monthly means).

Enter the number and name for the climate type and subtype in the blank spaces provided. These should follow the system given in the legend of the world climate map, text Figure 7.2.

If your instructor requires the Köppen climate system, identify the climate by use of the Köppen-Geiger world map, text p. 224–225, and text definitions and boundary graphs, p. 222–225. Enter the Köppen code symbol and Köppen name in the blank spaces provided beside each climograph.

Station Data

(a) Brisbane, Queensland, Australia 27½°S 153°E

	J	F	M	A	M	J	J	A	S	O	N	D	Mean
T (°F)	77	76.5	74	70	65	60	58.5	60.5	65.5	70	73	76	69
P (in.)	6.4	6.3	5.7	3.7	2.8	2.6	2.2	1.9	1.9	2.5	3.7	5.0	

(b) Zakinthos, Greece 38°N 21°E

	J	F	M	A	M	J	J	A	S	O	N	D	Mean
T (°F)	52	57.5	55.5	60.5	67.5	74.5	79.5	79.5	75.5	68.5	61	55.5	65
P (in.)	7.2	5.3	3.4	2.2	1.2	0.3	0.1	0.4	1.4	5.1	8.1	9.2	

(3) Oporto, Portugal 41°N 8½°W

	J	F	M	A	M	J	J	A	S	O	N	D	Mean
T (°F)	48	49.5	53	56	58.5	64.5	67	67.5	65.5	60	53.5	48.5	57.5
P (in.)	6.0	4.6	5.4	4.1	3.3	1.7	0.9	0.7	2.1	4.3	5.9	6.5	

(d) Volgograd (Stalingrad), Russia 49°N 44½°E

	J	F	M	A	M	J	J	A	S	O	N	D	Mean
T (°F)	9.5	15.5	25.5	44	60.5	69.5	74.5	72	59.5	45	30.5	20	43.5
P (in.)	0.9	1.0	0.6	0.6	1.0	1.9	0.9	0.8	0.7	1.0	1.5	1.3	

(e) Inch'on, Korea 27½°N 127°E

	J	F	M	A	M	J	J	A	S	O	N	D	Mean
T (°F)	26	29	38	50	59.5	68.6	75	77.5	68.5	58	43.5	30.5	52
P (in.)	0.8	0.7	1.2	2.6	3.3	3.9	10.9	8.8	4.3	1.6	1.6	1.1	

(f) Vardo, Norway, 70½°N 31°E

	J	F	M	A	M	J	J	A	S	O	N	D	Mean
T (°F)	23	22	24.5	30	36	42.5	48.5	48.5	43.5	35	29.5	26	34
P (in.)	2.5	2.5	2.3	1.5	1.3	1.3	1.5	1.7	1.9	2.5	2.1	2.4	

(g) Point Barrow, Alaska 71°N 157°W

	J	F	M	A	M	J	J	A	S	O	N	D	Mean
T (°F)	−15.5	−18.5	−15	−0.5	18.5	34	39.5	38.5	30.5	17	1	−10.5	10
P (in.)	0.2	0.1	0.1	0.1	0.1	0.3	0.9	0.8	0.5	0.5	0.3	0.2	

NAME _____   DATE _____

CLIMOGRAPHS

(a) Brisbane, Australia

Mean Ann. Temp.   _____

Ann. Temp. Ra.   _____

Ann. Precip.   _____

Climate number   _____

Climate name _____

_____

Köppen code   _____

Köppen name _____

_____

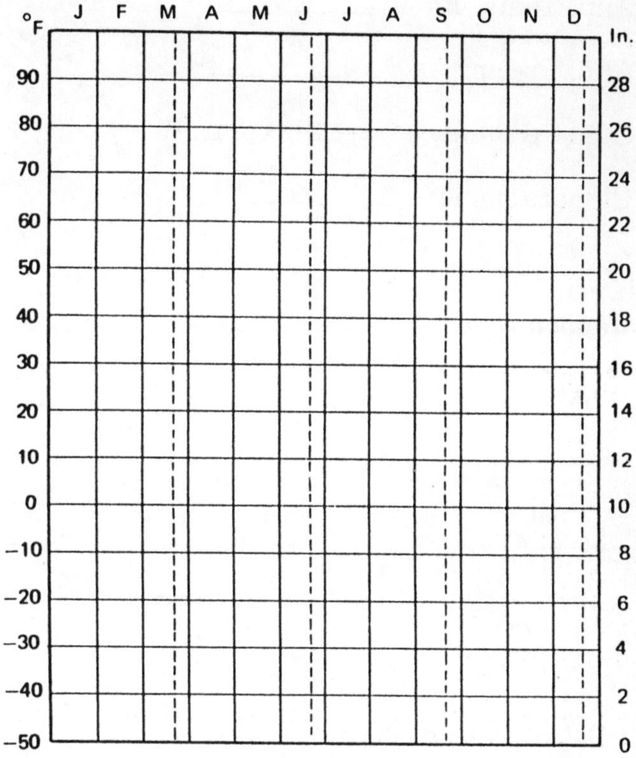

(b) Zakinthos, Greece

Mean Ann. Temp.   _____

Ann. Temp. Ra.   _____

Ann. Precip.   _____

Climate number   _____

Climate name _____

_____

Köppen code   _____

Köppen name _____

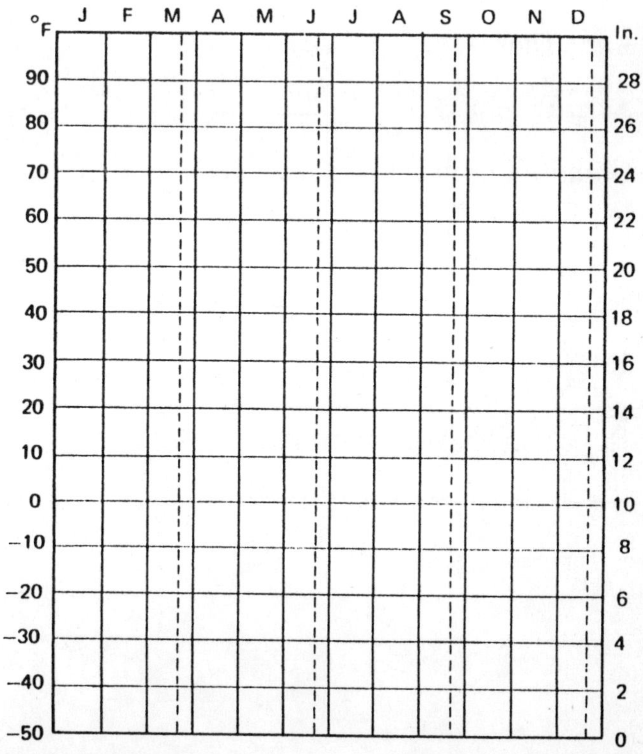

_____

(c) Oporto, Portugal

Mean Ann. Temp. _____

Ann. Temp. Ra. _____

Ann. Precip. _____

Climate number _____

Climate name _____

_____

Köppen code _____

Köppen name _____

_____

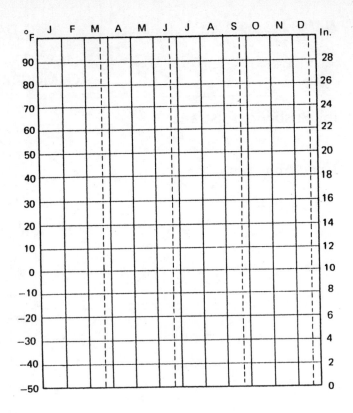

(d) Volgograd, Russia

Mean Ann. Temp. _____

Ann. Temp. Ra. _____

Ann. Precip. _____

Climate number _____

Climate name _____

_____

Köppen code _____

Köppen name _____

_____

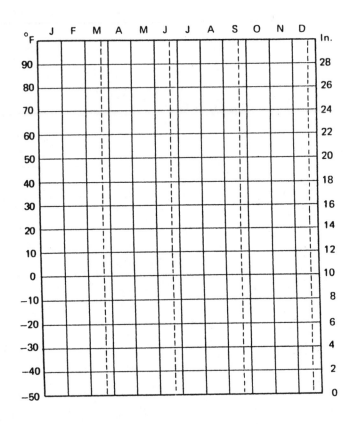

(e) Inch'on, Korea

Mean Ann. Temp. _____

Ann. Temp. Ra. _____

Ann. Precip. _____

Climate number _____

Climate name _____

_____

Köppen code _____

Köppen name _____

_____

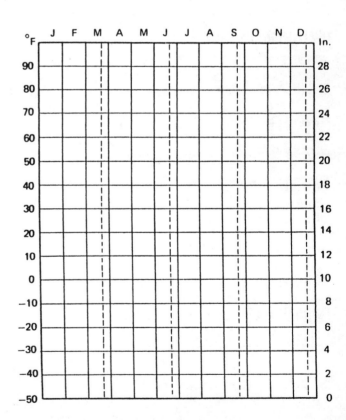

(f) Vardo, Norway

Mean Ann. Temp. _____

Ann. Temp. Ra. _____

Ann. Precip. _____

Climate number _____

Climate name _____

_____

Köppen code _____

Köppen name _____

_____

(g) Port Barrow, Alaska

Mean Ann. Temp. _____

Ann. Temp. Ra. _____

Ann. Precip. _____

Climate number _____

Climate name _____

_____

Köppen code _____

Köppen name _____

_____

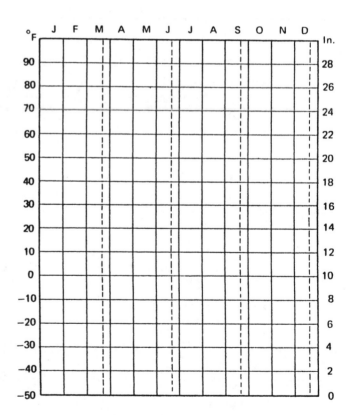

(Extra climograph)

Mean Ann. Temp. _____

Ann. Temp. Ra. _____

Ann. Precip. _____

Climate number _____

Climate name _____

_____

Köppen code _____

Köppen name _____

_____

QUESTIONS

(a) Brisbane, Australia

Brisbane, capital city of Queensland, is located on Moreton Bay on the east coast of Australia. Founded in 1828 as a penal colony, it is now an industrial city that exports wool, meat, fruit, and sugar, along with coal and other minerals brought from the arid hinterland lying west of the Great Dividing Range and the Queensland coast that stretches far to the north into the tropical zone.

(1) Compare the climograph of Brisbane with that of Charleston, South Carolina, text Figure 7.24, as to the thermal regimes shown.

_____

_____

_____

_____

_____

_____

(2) Compare the precipitation cycles of Brisbane and Charleston.

_____

_____

_____

_____

_____

(b) Zakinthos, Greece

Zakinthos is the southernmost of the Ionian Islands, lying off the west coast of Greece. Zakinthos is also the name of the island's chief town and port. According to legendary history of ancient Greece, Zacynthus, son of the Arcadian chief Dardanus, led the first settlers to this island from the heart of the Peloponnesus. Intensively cultivated, the island produces wine, currants, citrus fruits and olive oil—a typically Mediterranean fare.

(1) Compare the climograph of Zakinthos with that of Monterey, California, text Figure 7.23, as to the temperature cycle. Explain the differences.

_____

_____

_____

_____

_____

(2) Compare the precipitation regime of Zakinthos with that of Monterey. Explain the differences.

_____

_____

_____

_____

_____

_____

(c) Oporto, Portugal

Oporto (*Pôrto*), Portugal's second largest city, lies near the Atlantic at the mouth of the Douro River. The city is internationally known for its association with port wine—its most popular export. Oporto's long and checkered history includes occupation by the Moors between 716 and 1092. Climatologically speaking, this station is interesting as being transitional between the marine west-coast climate and the Mediterranean climate.

(1) Compare the climograph of Oporto with that of Vancouver, British Columbia, text Figure 7.29, as to the annual temperature cycle. Explain the differences.

_____

_____

_____

_____

(2) Compare the precipitation regimes of Oporto and Vancouver. Explain any differences you may find.

_____

_____

_____

_____

_____

(3) Compare the precipitation cycle of Oporto with that of a Mediterranean climate station, such as Zakinthos or Monterey. What is the essential difference?

_____

_____

_____

_____

_____

(d) Volgograd, U.S.S.R.

Volgograd lies on the west bank of the mighty Volga River in the midst of a vast plain that extends far eastward to the Kirgiz Steppe. Named Stalingrad in 1925, the city was nearly destroyed in World War II when under seige by Hitler's armies. The city's name was changed to Volgograd in 1961, under the regime of Nikita Khruschev, who had denounced Stalin and his dictatorship. Today a city of heavy industry, Volgograd derives hydroelectric power from an enormous dam that spans the Volga just above the city.

(1) Identify the climate and subtype for Volgograd. Compare the climograph with that of Pueblo, Colorado, text Figure 7.32, with respect to the temperature cycle. Explain the important differences.

_____

_____

_____

_____

_____

(2) Compare the precipitation regimes of Volvograd and Pueblo. Explain the important difference in the cycles.

_____

_____

_____

_____

_____

(e) Inch'on, Korea

Inch'on, a west-coast port city for South Korea's capital, Seoul, provides access to the shallow Yellow Sea, but is an industrial city in its own right. During the Korean War, Inch'on was the site of a daring landing of U.S. troops in September, 1950, executed under the command of General Douglas MacArthur to relieve enemy pressure on beleagured UN troops within the Pusan perimeter far to the south.

(1) Fans of TV's indestructable saga, MASH, have experienced the intense continentality of Korea's climate, including vivid episodes of summer heat wave and winter deep-freeze. Compare the climographs of Inch'on and Madison, Wisconsin, text Figure 7.33, as to the annual temperature cycle. Explain the important differences.

_____

_____

_____

(2) Veteran MASH viewers notice that rain is almost never shown, perhaps because the filming was done only on dry days in southern California's Santa Monica Mountains. Compare the precipitation cycles of Inch'on and Madison. Explain the significant difference.

_____

_____

_____

(f) Vardo, Norway

Vardo is a small port town on the Arctic Ocean lying 4 degrees of latitude north of the arctic circle, equivalent to about 280 miles. Despite this arctic location, the port is ice-free, thanks to the mild waters of the North Atlantic drift that sweep eastward around the North Cape. History has it that in 1320 the northern most fortress in the world was built at Vardo, which thereafter enjoyed brisk trade with Russia and Finland. The explorer Fridtjof Nansen used Vardo as his base of operation for arctic expeditions in the 1890s.

(1) Placing Vardo's climate in one of the standard types described in your textbook poses some problems. Careful analysis of the temperature data is needed, along with comparisons with both the boreal forest climate (11) illustrated by the climograph for Ft. Vermilion, Alberta (text Figure 7.38), and the tundra climate (12) shown in the climograph for Upernivik, Greenland (Figure 7.41) and that for Point Barrow in this exercise. Make these comparisons in terms of (a) annual temperature range and (b) mean annual temperature. Explain the similarities and differences.

_____

_____

_____

_____

_____

_____

_____

_____

(2) Make the same climograph comparisons with respect to total annual precipitation. Draw conclusions as to the meaning of the data.

_____

_____

_____

_____

_____

(3) Explain the exceptionally high precipitation of Vardo as compared with the other stations examined above.

_____

_____

_____

_____

(g) Point Barrow, Alaska

Point Barrow, on the shores of the Arctic Ocean, is the most northerly point of Alaska; it lies at latitude 71°N, which is about 5° north of the arctic circle (equivalent to about 330 mi). Nearby is the small city of Barrow, once a rather insignificant but ancient Eskimo settlement. After World War II, the U.S. Navy placed an arctic research station in Barrow, and the U.S. Air Force set up a radar station close by as part of the Distant Early Warning Line. Wiley Post, a pioneer American aviator, used Point Barrow as a stopover on his arctic round-the-world flight in 1931, but later he and a distinguished passenger, humorist Will Rogers died in a plane crash near Point Barrow.

(1) Barrow's temperature cycle was noted in the previous exercise as having a much greater annual range than Vardo, although both are coastal stations on the Arctic Ocean. How do you explain this striking difference?

_____

_____

_____

_____

_____

(2) Explain the sharp increase in precipitation in July and August at Barrow.

_____

_____

_____

_____

NAME _____   DATE _____

Exercise 7-E Identifying Köppen Climates

[Text p. 222–225.]

Table A gives climate data for ten stations, designated by letters (a) through (j). Monthly mean temperatures (°F) and mean annual temperature are given in the upper line of figures. The lower line gives mean monthly precipitation in inches, and mean annual total.

Table A

		J	F	M	A	M	J	J	A	S	O	N	D	Year
(a)	°F	13.3	13.6	27.0	43.9	53.8	62.2	69.1	65.8	56.1	43.9	30.9	20.1	41.5
	In.	0.7	0.6	0.5	1.0	2.0	2.9	1.8	1.2	1.3	0.7	0.6	0.6	13.9
(b)	°F	−14.4	7.4	13.3	33.8	46.8	58.5	62.6	58.5	46.0	30.0	6.8	−9.2	27.1
	In.	0.1	0.1	0.1	0.4	0.9	1.7	2.4	3.0	1.3	0.3	0.2	0.2	10.3
(c)	°F	71.1	73.6	83.1	91.6	94.5	93.7	89.2	86.5	89.2	88.9	80.8	71.1	84.4
	In.	0	0	0.1	0	0.3	0.9	3.5	2.8	1.1	0.4	0	0	9.0
(d)	°F	33.4	35.2	41.9	53.2	64.0	72.1	76.6	75.0	68.0	57.7	46.2	36.9	55.0
	In.	3.4	3.7	3.6	3.4	3.5	3.9	4.6	4.3	3.4	2.8	2.6	3.3	42.5
(e)	°F	47.8	49.5	52.5	58.3	66.4	74.1	79.7	79.5	73.2	66.2	57.0	57.8	63.0
	In.	2.2	1.8	1.3	0.9	0.8	0.6	0.3	0.6	0.7	1.4	2.9	2.5	16.0
(f)	°F	77.9	78.4	79.3	79.9	80.6	79.9	80.2	79.7	79.5	79.7	79.0	78.3	79.3
	In.	9.7	7.1	7.3	7.8	6.5	7.0	6.7	7.8	6.9	7.9	10.1	10.5	95.1
(g)	°F	80.8	80.6	80.8	80.4	78.4	75.0	74.8	78.1	81.3	82.6	81.7	81.0	79.7
	In.	9.6	8.9	8.1	4.1	2.0	0.2	0.1	1.1	2.0	4.4	6.0	7.9	54.5
(h)	°F	−14.3	−12.8	−2.0	12.0	28.6	38.7	45.7	44.2	36.3	26.1	12.4	−3.8	17.6
	In.	1.1	1.3	0.8	1.5	1.5	1.1	2.6	2.0	1.2	1.5	2.2	1.5	18.3
(i)	°F	63.3	63.0	62.2	58.3	54.1	51.6	50.4	53.4	60.1	63.5	64.6	63.9	59.0
	In.	10.6	8.4	8.3	2.2	0.9	0.6	0.8	0.3	0.3	1.3	3.5	8.5	45.7
(j)	°F	22.6	23.9	31.6	42.4	52.5	61.9	67.5	65.7	59.2	49.6	37.9	27.7	45.1
	In.	3.9	4.3	3.8	3.4	3.3	3.3	3.2	3.1	3.1	3.1	3.5	3.9	41.9

Determine the Köppen climate to which each station belongs, referring to the climate list and symbols on text p. 222–225. Enter the code symbols in the spaces provided. If possible, give the third letter of the code as well. For each, state the criteria used in assigning the station to that climate.

(a) Code _____ Criteria _____

_____

_____

_____

(b) Code _____ Criteria _____

_____

_____

_____

(c) Code _____ Criteria _____

_____

_____

_____

(d) Code _____ Criteria _____

_____

_____

_____

(e) Code _____ Criteria _____

_____

_____

_____

(f) Code _____ Criteria _____

_____

_____

_____

(g) Code _____ Criteria _____

_____

_____

_____

(h) Code _____ Criteria _____

_____

_____

_____

(i) Code _____ Criteria _____

_____

_____

_____

(j) Code _____ Criteria _____

_____

_____

_____

NAME _____   DATE _____

Exercise 8-A Energy Flow in Ecosystems

[Text p. 271–272, Figure 8.2.]

The *food web*, or *food chain*, as an energy system is explained on pages 271 and 272 of your textbook. Here, we develop further the passage of energy upward in the food chain, shown in Figure 8.2. This example applies to a terrestrial ecosystem in which 10% of the available energy is passed along the food chain to the next level. The primary production rate is taken as 18,000 kilocalories per square meter per year (Kcal/m^2/yr).

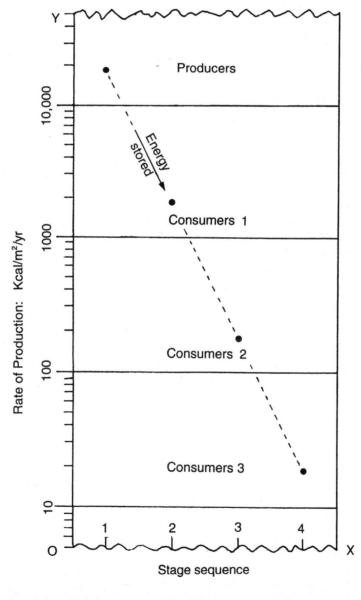

Graph A

EX. 8-A

We have reproduced the data in Figure 8.2 as Graph A, which uses a black dot for each of the four trophic levels. We use a logarithmic energy-change scale on the ordinate (vertical scale) and an arithmetic scale as the abscissa (horizontal scale). The four classes of organisms (producers and three layers of consumers) are designated on the abscissa by a number sequence 1 through 4. Thus, the flow of energy appears as a succession of downward steps, as the trophic level rises.

(1) Uppermost dot represents the level of energy stored in the system by the action of primary producers. What is the origin of this stored energy and by what biochemical process is it produced?

_____

_____

_____

(2) When only 10 percent of the system energy is passed on to the next higher level, what happens to the remaining 90 percent? Does it leave the system, and if so, how does this happen?

_____

_____

_____

(3) On Graph A, insert a set of four points to show the 10 percent energy loss (or remaining 90 percent) that corresponds with each stage.

Graph B is presented as a blank graph on which both axes are on arithmetic scales. Plot the four points for the 10-percent program of energy storage reduction.

(4) Describe the curve you plotted on Graph B. What meaning attaches to the shape of the curve?

_____

_____

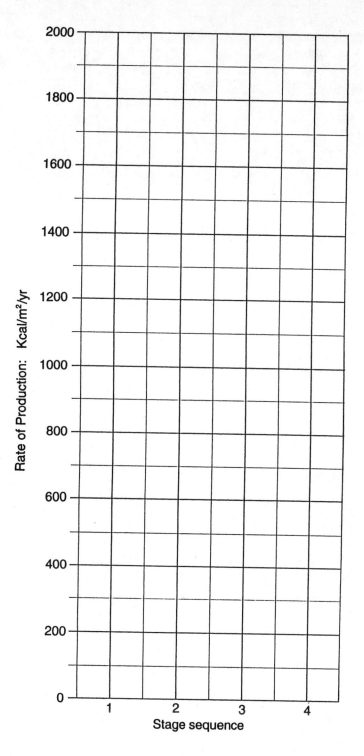

Graph B

NAME _____   DATE _____

Exercise 8-B Photosynthesis and Carbon Dioxide Concentration

[Text p. 272–273, Figures 8.3–8.5.]

The possible global effects of an increased content of atmospheric carbon dioxide (CO_2) is a persistent theme throughout your textbook. The effect of higher levels of CO_2 on the growth of green plants has been studied in botanical laboratories to determine the rates of photosynthesis associated with controlled CO_2 levels in the surrounding atmosphere.

Graph A displays the results of a study in which wheat seedlings were exposed to four different levels of illumination. Rate of photosynthesis is labeled on the abscissa in *milliliters of CO_2 evolved per minute*. Intensified light levels (labeled in *foot-candles*) gave consistently greater rates of photosynthesis.

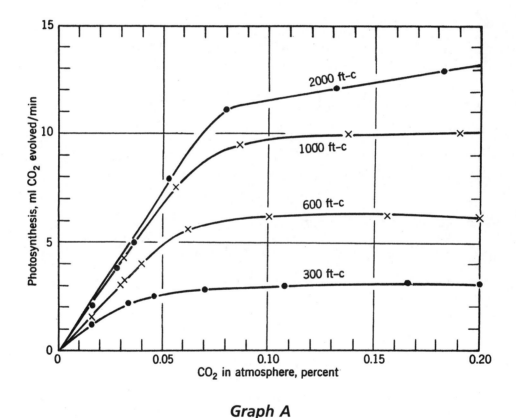

Graph A

(1) Locate on the baseline of Graph A the present global value of atmospheric carbon dioxide.

Enter present value of CO_2: _____ .

(2) Make a general statement describing the plotted curves. What change of form takes place as percent of CO_2 is traced from left to right?

_____

_____

_____

(3) Offer a physical explanation for the typical curve you have described in Question 2.

_____

_____

_____

_____

(4) Relate the curve form you have described to the possible global warming trend that may have already begun in response to the observed increase in atmospheric CO_2.

_____

_____

_____

Table A _Values of CO$_2$ Concentration_

Hours	12m	3	6	9	12n	3	6	9	12m
Height:									
1 m:	347	352	350	334	316	316	325	342	345
10 m:	329	333	332	315	304	303	309	316	327
20 m:	322	332	330	311	307	307	308	313	322

Based on data from B. Bolin, 1970, _Scientific American_, vol. 233, no. 3, p. 127.

The Daily Cycle of Carbon Dioxide

Besides the fact that CO_2 concentrations are normally limiting to the rate of photosynthesis, another important implication of the generally low level of CO_2 in the atmosphere is the local removal of CO_2 from the atmosphere by photosynthesis concentrations. Table A lists CO_2 concentrations measured throughout a full day at three levels above ground in a dense forest. Measurements were taken at heights of 1, 10, and 20 meters above the ground. Plot each observation on blank Graph B and connect each sequence of values with straight line segments.

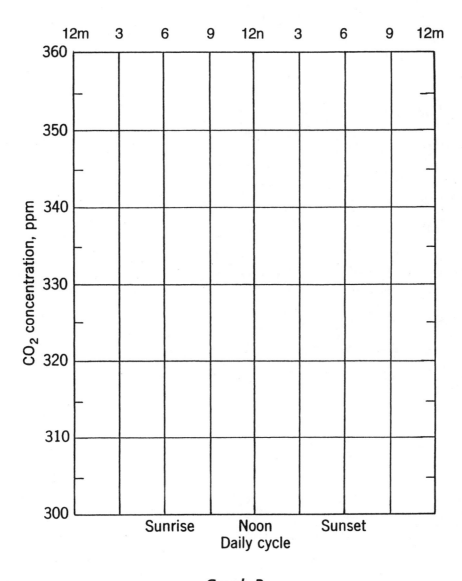

Graph B

(5) Describe the rhythm of rise and fall of CO_2 concentration shown on the graph. Offer an explanation for this rhythm.

_____

_____

_____

(6) Why does the 1-m cycle show a greater concentration throughout the cycle than at the 10 m and 20 m levels?

_____

_____

_____

Copyright © by Arthur N. Strahler

NAME _____   DATE _____

Exercise 9-A The Forest Biome

[Text p. 315–328, Figure 9.3.]

Scarcely a week goes by without news media mention of the rapid destruction of the low-latitude rainforests and their unique ecosystem. Environmental activists among the geographers and ecologists in the highly developed nations of the Anglo-European culture cry out against this devastation, seeing it as an environmental disaster. All well and good, perhaps, but managers of those developing nations where the rainforests stood sometimes see the issue in a different light. They point out that Europeans, whose rapidly expanding population desperately needed food, centuries ago destroyed their own native forests to place the land in cultivation. Under continued pressure, emigrant Europeans then settled North America, destroying the forests and prairies. Have we forgotten the loss of those primeval forest and prairie ecosystems? From Brazil to Indonesia hostility greets the environmentalists who would save the rainforest. Have you a practical solution to this deep-seated conflict?

Physical geography can be thought of as having layers of knowledge, stacked one upon another. One such stack consists of world climates, at the base, world soils upon that, and world vegetation as the top layer. (Some instructors prefer to cover vegetation before soils.) These layers interact, of course, and the idea is more an abstraction than a reality, but it gives order to the way physical geography can be studied. Now, we turn to the relationship of world vegetation to both soils and climate. Taking on this task is made easier when we concentrate on one biome at a time, and on the formation classes within a biome, one at a time. For information on soil orders and suborders, you will need to look ahead into Chapter 10.

Map B (end of exercise) shows details of the world distribution of the formation classes. Refer to it when you need information not available on your textbook map of world vegetation, Figure 9.3. You will also need to use maps A, and B of Exercise 10-B.

Low-Latitude Rainforest

(1) Enter the climates and soils associated with the low-latitude rainforest (including equatorial rainforest and tropical rainforest):

Climate types and subtypes (number and name): _____

_____

Soil orders and suborders: _____

(2) From climograph figures in text Chapter 7, select those stations whose climates are linked closely to the low-latitude rainforest. Enter below the station name and country. Follow it with corresponding entries of climate type and soil order (or suborder). Locate and label each station on the world vegetation map, Map B of this exercise.

Station: _____   Climate: _____

    Soil: _____

Station: _____   Climate: _____

    Soil: _____

Station: _____   Climate: _____

    Soil: _____

(3) Follow the same instructions as in Question 2 for stations shown by climographs of Exercise 7-B.

Station: _____   Climate: _____

    Soil: _____

Station: _____   Climate: _____

    Soil: _____

Station: _____   Climate: _____

    Soil: _____

Monsoon Forest

Monsoon forests of India, Ceylon, Burma, Thailand, and Kampuchea (Cambodia) once held the great teakwood tree, sought after the world over for its fine lumber. Now, most of that resource is gone, and with it the Indian elephant, once widely domesticated in great numbers to handle the teakwood logs. The remaining domesticated elephants now have little to do but appear in festivals and public ceremonies. Those remaining in the wild state face extermination at the hands of hunters, despite local laws to curb poaching.

Monsoon forest, also known as rain-green forest, is a tropical-zone formation class developed in certain parts of southeast Asia, Central and South America, Africa, the East Indies, and Australia. On the world vegetation map, text Figure 9.3, it is included within a much larger class called tropical raingreen vegetation, most of which falls in the savanna biome as savanna woodland and thorntree tall-grass savanna. Find monsoon forest on Map B, designated as Fmo.

(4) Enter the climates and soils associated with the monsoon forest:

Climate types and subtypes: _____

_____

Soil orders and suborders: _____

_____

(5) From text Chapter 7 and Exercise 7-B select those stations whose climographs are linked closely to the monsoon forest. Enter the station name and country. For each, enter the appropriate options for (a), (b), and (c), above. Locate and label each station on the world vegetation map, Map B of this exercise.

Station: _____   Climate: _____

Soil: _____

Station: _____   Climate: _____

Soil: _____

Subtropical Evergreen Forests

The subtropical evergreen forests vary widely in composition from place to place in different world localities in which they occur. Some are composed broadleaved trees that hold their leaves through the season, others consist largely of needleleaf evergreen trees, particularly the pines, and there are also mixed broadleaf/needleleaf forests. The latitudinal span is mostly within the subtropical zone, but it extends farther poleward in the western Pacific Ocean in both Japan and New Zealand, showing the strong marine influence of that great ocean. Map B shows the distribution of the subtropical evergreen forest, Fbe.

(6) Enter the climates and soils of the subtropical evergreen forest:

Climate types and subtypes: _____

_____

Soil orders and suborders: _____

_____

(7) From text Chapter 7 and Exercise 7-D, select a station whose climograph is linked closely to the subtropical evergreen forest. Enter the station name and country. Follow it with corresponding entries of climate type and soil order (or suborder). Locate and label each station on the world vegetation map, Map B.

Station: _____   Climate: _____

Soil: _____

Station: _____   Climate: _____

Soil: _____

For formation classes in the U.S. and southern Canada, we supplement your textbook's world vegetation map with Map A, showing vegetation types of that region. It gives specific information on the composition of forest, grassland, and shrub formation classes.

(8) Examine Map A closely in southeastern region of the United States. On the world vegetation map, text Figure 9.3, this area shows subtropical evergreen forest, with an arrow reading "Southern pine forest." What kinds of forest vegetation are shown here in Map A? With what kind of land surface (upland or marsh land) is each associated?

_____

_____

_____

Midlatitude Deciduous Forests

The midlatitude decidious forests, often called "hardwood forests," once covered most of North America from the Atlantic coast to the Mississippi valley and from Ontario to Georgia. Much of that area is today cultivated farmland, with isolated timber plots. In contrast, the rugged upland surfaces of the Appalachian mountains and higher plateaus maintain these forests, albeit mostly regrown since settlers cut over the climax forest they found as they spread westward through what was then called "the wilderness."

(9) Enter the climates and soils associated with the midlatitude deciduous forest:

Climate types and subtypes: _____

_____

Soil orders and suborders: _____

_____

(10) From text Chapter 7 and Exercise 7-D, select climographs that are linked closely to the midlatitude deciduous forest. Enter the station name. Follow it with corresponding entries of the associated climate type and soil order. Locate and label the station on Map A.

Station: _____   Climate: _____

     Soil: _____

Station: _____   Climate: _____

     Soil: _____

(11) On Map A, what classes of midlatitude deciduous forests are recognized?

_____

_____

_____

_____

Cold Needleleaf Forests

The cold needleleaf forests occur almost exclusively in the northern hemisphere, where they occupy an enormous belt across North America and Eurasia. In the southern hemisphere, they occur only in a narrow coastal zone of southernmost South America. To these are added the high-altitude needleleaf forests of mountainous western North America and Europe. Map B shows four formation classes of the cold needleleaf forest: Fbo, Fbd, Fbl, Fl.

(12) Enter the climates and soils associated with the cold needleleaf forests:

    Climate types and subtypes: _____

_____

    Soil orders and suborders: _____

_____

(13) From text Chapter 7 and Exercise 7-D, select those stations whose climographs are linked closely to the cold needleleaf forest. Enter the station name and country. Follow it with corresponding entries of the associated climate type and soil order. Locate and label each station on the world vegetation map, Map B of this exercise.

Station: _____   Climate: _____

Soil: _____

Station: _____   Climate: _____

Soil: _____

(14) Refer to the eastern half of Map A. (a) What name is given to the cold needleleaf forest along the northern border of the map? (b) To the needleleaf forest in the Great Lakes area (Minnesota, northern Wisconsin, peninsular Michigan)?

(a) _____

(b) _____

Note: Your world vegetation map, text Figure 9.3, shows much of the area of BM (northeastern hardwoods) of Figure A as being in the cold needleleaf forests. This is because the BM class is mixed deciduous/hardwood forest, in which needleleaf trees (hemlock and others) are locally important or dominant, especially in sandy soils (Spodosls) and at higher altitudes.

(15) On Map A, what kinds of cold needleleaf forest are shown in the Sierras, Cascades, and Rockies? Give the locations of these forest types.

_____

_____

_____

_____

_____

_____

_____

Coastal Forest

In the Pacific Northwest of the United States, the great groves of ancient old-growth red-woods and Douglas fir that remain available to logging are rapidly disappearing. Are we carrying out in the Pacific Northwest the same act of ecosystem destruction many of us so strongly deplore when it's the Brazilians who are doing it in Amazonia? In an attempt to halt the cutting of an area of old-growth Douglas fir forest on federal land in the Olympic Peninsula, the U.S. Forest service proposed to take a large area out of production in order to save the spotted owl, an endangered species. The local loggers strongly objected, for it would have cost them their jobs. An emotional debate ensued between loggers and environmentalists. Where did you stand on this issue? Did you carry a bumper sticker reading *Save a logger—kill a spotted owl*?

The coastal forest formation class is given special treatment in the mountainous Pacific coast ranges from northern California, through Oregon, Washington, and British Columbia to southern Alaska, in the latitude range 40°–60°N. A needleleaf forest, its composition changes from south to north. Redwood is important in California, giving way to Douglas fir in Oregon and Washington, then to spruce in the northern section.

(16) Enter the climates and soils associated with the coastal forest:

Climate types and subtypes: _____

_____

Soil orders and suborders: _____

(17) From textbook Chapter 10 select a station whose climograph is linked closely to the coastal forest. Enter the station name and location. Follow it with corresponding entries of climate type and soil. Locate and label each station on the world vegetation map, Map B, and on Map A.

Station: _____   Climate: _____

Soil: _____

Note that on the world vegetation map, Map B, the area of coastal forest is extended inland in a long tongue that reaches to the Rockies. Examine this entire area on Map A.

(18) What coastal forest types are included under the pattern of bold horizontal lines on map A? Give the name and location of each subtype.

_____

_____

_____

Sclerophyll Forest

Sclerophyll forest is an open forest of the humid and subhumid subtypes of the Mediterranean climate (7h, 7sh). On the world vegetation map, text Figure 9.3, it is lumped together with Mediterranean woodland and scrub formations as sclerophyllous vegetation. Map B shows formation classes of the sclerophyllous vegetation: Fsm, Fss, Fsa, Ssa. For a true example of the forest formation, we would need to turn to the Australian eucalyptus forests (Fsa). These occur not only in the Perth region of southwestern Western Australia (7h, 7sh) and the Adelaide region of South Australia (7s), but also on the eastern continental margin in the Australian Alps, Blue Mountains, and New England Range of Victoria and New South Wales in other climates (5s, 6). Although these eucalyptus forests in their original climax form consisted of trees of great height, their crowns were small and provided little shade to the ground. A ground cover of grasses and shrubs was typical of these forests.

The sclerophyllous chaparral of central and southern California has been called a "dwarf forest." This designation applies to what are dense stands of dwarf trees that shade a large proportion of the ground. Much of it is better described as woodland. This formation class (Fss on Map B) is described in your text, p. 328 and Figure 9.18.

(19) On Map A, where is chaparral shown, and what name is given to it?

_____

_____

_____

_____

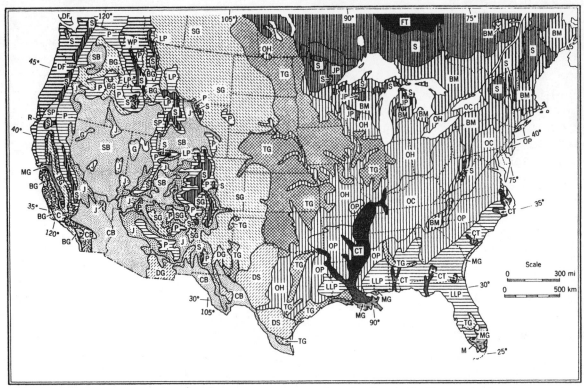

EASTERN FOREST VEGETATION

FT   Subarctic forest-tundra transition (Canada)
S   Spruce-fir (Northern coniferous forest)
JP   Jack, red, and white pines (Northeastern pine forest)
BM   Birch-beech-maple-hemlock (Northeastern hardwoods)
Oak forest (Southern hardwood forest):
    OC   Chestnut-chestnut oak-yellow poplar
    OH   Oak-hickory
    OP   Oak-pine
CT   Cypress-tupelo-red gum (River bottom forest)
LLP   Longleaf-loblolly-slash pines (Southeastern pine forest)
M   Mangrove (Subtropical forest)

WESTERN FOREST VEGETATION

S   Spruce-fir (Northern coniferous forest)
Cedar-hemlock (Northwestern coniferous forest):
    WP   Western larch-western white pine
    DF   Pacific Douglas fir
    R   Redwood

Yellow pine-Douglas fir (Western pine forest)
    SP   Yellow pine-sugar pine
    P   Yellow pine-Douglas fir
    LP   Lodgepole pine
J   Pinon-Juniper (Southwestern coniferous woodland)
C   Chaparral (Southwestern broad-leaved woodland)

DESERT SHRUB VEGETATION

SB   Sagebrush (Northern desert shrub)
CB   Creosote bush (Southern desert shrub)
G   Greasewood (Salt desert shrub)

GRASS VEGETATION

TG   Tall grass (Prairie grassland)
SG   Short grass (Plains grassland)
DG   Mesquite-grass (Desert grassland)
DS   Mesquite and desert grass savanna (Desert savanna)
BG   Bunch grass (Pacific grassland)
MG   Marsh grass (Marsh grassland)
Alpine meadow (Not shown)

Map A *Vegetation of the contiguous 48 United States and southern Canada. (Based on maps of H.L. Shantz and Raphael Zon in* **Atlas of American Agriculture**; *Canada Department of Forestry, Bulletin 12.)*

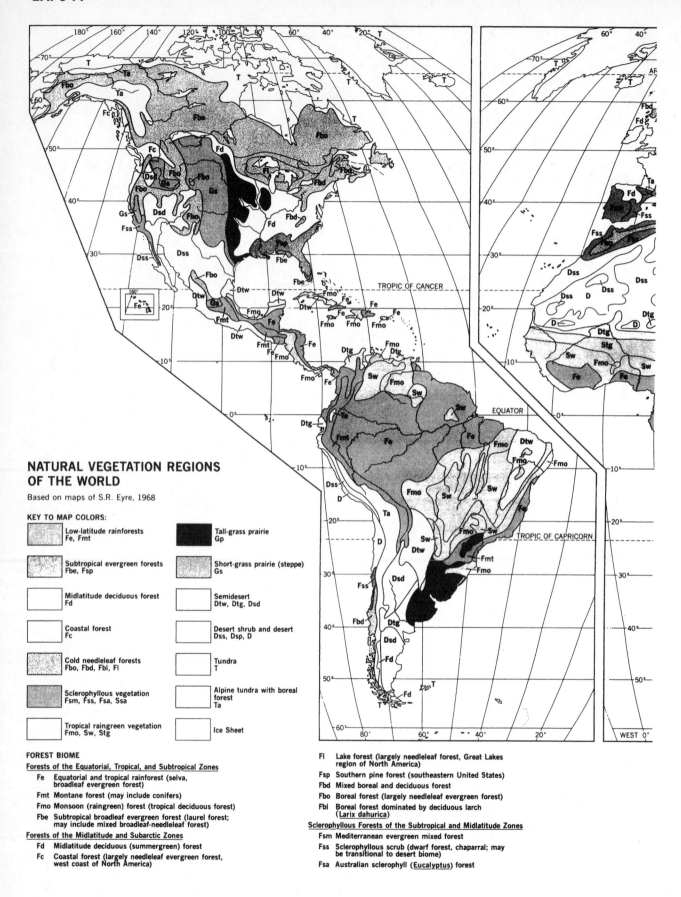

NATURAL VEGETATION REGIONS OF THE WORLD

Based on maps of S.R. Eyre, 1968

KEY TO MAP COLORS:

Low-latitude rainforests
Fe, Fmt

Subtropical evergreen forests
Fbe, Fsp

Midlatitude deciduous forest
Fd

Coastal forest
Fc

Cold needleleaf forests
Fbo, Fbd, Fbl, Fl

Sclerophyllous vegetation
Fsm, Fss, Fsa, Ssa

Tropical raingreen vegetation
Fmo, Sw, Stg

Tall-grass prairie
Gp

Short-grass prairie (steppe)
Gs

Semidesert
Dtw, Dtg, Dsd

Desert shrub and desert
Dss, Dsp, D

Tundra
T

Alpine tundra with boreal forest
Ta

Ice Sheet

FOREST BIOME

Forests of the Equatorial, Tropical, and Subtropical Zones

- Fe   Equatorial and tropical rainforest (selva, broadleaf evergreen forest)
- Fmt Montane forest (may include conifers)
- Fmo Monsoon (raingreen) forest (tropical deciduous forest)
- Fbe Subtropical broadleaf evergreen forest (laurel forest; may include mixed broadleaf-needleleaf forest)

Forests of the Midlatitude and Subarctic Zones

- Fd   Midlatitude deciduous (summergreen) forest
- Fc   Coastal forest (largely needleleaf evergreen forest, west coast of North America)

- Fl   Lake forest (largely needleleaf forest, Great Lakes region of North America)
- Fsp Southern pine forest (southeastern United States)
- Fbd Mixed boreal and deciduous forest
- Fbo Boreal forest (largely needleleaf evergreen forest)
- Fbl Boreal forest dominated by deciduous larch (*Larix dahurica*)

Sclerophyllous Forests of the Subtropical and Midlatitude Zones

- Fsm Mediterranean evergreen mixed forest
- Fss Sclerophyllous scrub (dwarf forest, chaparral; may be transitional to desert biome)
- Fsa Australian sclerophyll (*Eucalyptus*) forest

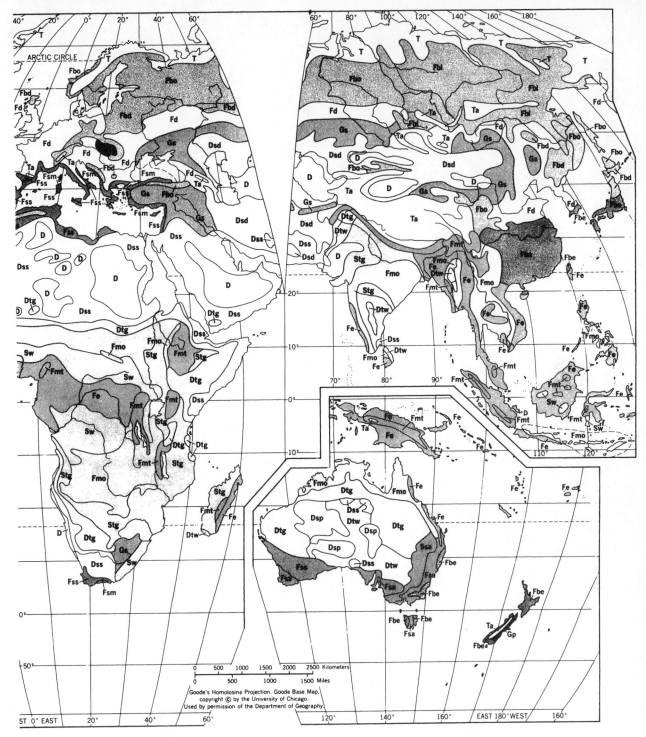

Goode's Homolosine Projection. Goode Base Map,
copyright © by the University of Chicago.
Used by permission of the Department of Geography.

SAVANNA BIOME
- Sw Savanna woodland (broadleaf tree savanna)
- Stg Thorntree-tall grass savanna
- Ssa Australian sclerophyllous tree savanna

GRASSLAND BIOME
- Gp Tall-grass prairie
- Gs Short-grass prairie (steppe)

DESERT BIOME
- Dtw Thorn forest and thorn woodland (may be transitional to forest)
- Dtg Thorntree-desert grass savanna

- Dsd Semidesert scrub and woodland
- Dss Semidesert scrub
- Dsp Desert alternating with porcupine grass semidesert
- D Desert

TUNDRA BIOME
- T Arctic tundra
- Ta Alpine tundra (includes boreal forest)

Data source: S. R. Eyre, Vegetation and soils; a world picture,
Second Edition, Aldine Publishing Company, copyright © 1968 by S. R. Eyre.
See Appendix I, Maps 1–10. Map boundaries and classes have been
simplified and modified by the authors with permission of S. R. Eyre,
Edward Arnold (Publishers) Ltd., and The Aldine Publishing Company.

NAME _____   DATE _____

Exercise 9-B Savanna, Grassland, Desert, and Tundra Biomes

[Text p. 328–335, Figure 9.3.]

Following the pattern of Exercise 9-A, we continue to investigate the relationships of biomes and formation classes to soils and climate. (Note: Maps A and B are found in Exercise 9-A.)

Savanna Biome

The savanna biome lies in the tropical zone, transitional between the low-latitude rainforests and the tropical deserts. For the masses of humanity that populate the savanna biome in South America, Africa, India, and Southeast Asia, this transitional zone offers only limited opportunities to provide adequate food without extensive irrigation projects. Evolution by natural selection through millions of years endowed the ecosystems of this biome with a remarkable ability to flourish in a stressful regime of intense seasonal drought. Cultural evolution over many millennia produced an equally remarkable adaptation of humans to the climate, soils, and vegetation of this biome. That long era of a delicately balanced ecology encompassing plants and animals seems now to be ending under the impact of other humans culturally adapted to a midlatitude environment and choosing to force that alien culture upon a very unfamiliar ecosystem. The better we understand the environmental science of this biome, the better will we be able to understand what went wrong over the past two centuries of European influence and what solutions can alleviate, rather than intensify, problems already out of hand.

(1) Enter the climates and soils associated with the savanna biome, including both savanna woodland (Sw) and thorntree/tall-grass savanna (Stg) formation classes:

Climate types and subtypes (number and name): _____

_____

_____

(2) On the world vegetation map, Map B of Exercise 9-A, Color green all areas of Savanna woodland (Sw). Color red all areas of Thorntree tall-grass savanna (Stg). Compare the latitude range of Sw in Africa with that in South America. Where is Stg found at its highest latitude?

Africa: Max. N. lat.: _____.   Max. S. latitude: _____   Range: _____ degrees.

S. Amer.: Max. N. lat.: _____   Max. S. latitude _____   Range: _____ degrees.

Highest latitude of Stg: _____

(3) From text Chapter 9 and Exercise 9-D, select those tropical stations whose climographs are closely linked to the savanna woodland formation class (Sw). Enter the station name and country. Follow it with corresponding entries of climate type and soil order (or suborder). Locate and label each station on the world vegetation map, Map B of Exercise 9-A.

Station: _____   Climate: _____

Soil: _____

Station: _____   Climate: _____

Soil: _____

Grassland Biome

We must treat separately the two major formation classes of the grassland biome: tall-grass prairie (Gp) and steppe, or short-grass prairie (Gs). Each occupies a distinctively different climatic zone and soil suborder. On your text map, Figure 9.3, the two formation classes are shown in different colors.

(4) Enter the climate and soil associated with the tall-grass prairie:

Climate type and subtype (number and name): _____

Soil orders and suborders: _____

(5) On Map A (Exercise 8-A), large areas of tall-grass prairie (TG) are shown

(a) List all of the states and provinces in which tall-grass prairie (TG) is present.

_____

_____

(b) In Canada and the continental U.S., the longitudinal extent of tall-grass prairie

ranges from about _____ to _____.

(c) In latitudinal extent, from about _____ to _____.

(6) We can use the state of Iowa as a good representative of the tall-grass prairie region. Locate this state on Map A. In what climate types does it lie? What soil suborder is present here?

Climate: _____

Soil order and suborder: _____

Copyright © by Arthur N. Strahler

(7) On your world vegetation map, text Figure 9.3, a large area of tall-grass prairie is shown over parts of Argentina and Uruguay. This region is known as the *Pampa*. What climate types and subtypes are represented here? What soil order is present?

Climate: _____

Soil order: _____

Steppe, or short-grass prairie (Gs), occupies large areas of both North America and Eurasia. Here, we concentrate on the North American steppes, shown on Map A of Exercise 9-A. Use also the U.S.-Canada soils map, Map A of Exercise 10-B.

(8) With what climate and climate subtypes is the North American steppe grassland identified? With what soil suborders?

Climate: _____

Soil suborders: _____

(9) From text Chapter 7 or Exercise 7-D, select a station whose climograph is closely linked to the steppe, or short-grass prairie. Enter below the station name and location. Follow with climate type and subtype and with soil order and suborder.

Station: _____

Climate: _____

Soil order and suborder: _____

Desert Biome

For the desert biome, we concentrate on the vegetation of deserts of the American west, where both semidesert and desert formation classes are recognized. Use vegetation Map A of Exercise 9-A, the U.S.-Canada maps of soils (Exercise 10-B, Map A), and the U.S. Canada climate map (Exercise 10-B, Map B).

(10) Where do we find the semidesert vegetation areas in the U.S., west of the Rockies? What climate subtypes are involved? (Use Map B. Exercise 9-A.)

_____

_____

_____

_____

_____

(11) What vegetation type is shown on Map A of Exercise 9-A for this northern semi-desert region? For the dry desert (5d) of the Sonora-Mojave desert that lies to the south?

_____

_____

_____

_____

(12) What soil orders correspond with the northern area? the southern area? (See Map A, Exercise 10-B)

    Northern area: _____

    _____

    Southern area: _____

Tundra Biome

(13) Climographs of the tundra climate (12) appear as text Figure 7.41, Upernivik, Greenland, and as Exercise 10-C (g), Point Barrow, Alaska. Locate and label these stations on the world vegetation map, Map B of Exercise 9-A.

(14) For about how many consecutive months of the year is soil water solidly frozen in the arctic tundra at each of these three stations?

    Upernivik: _____ mo. Point Barrow: _____ mo

(15) What soil order, suborder, and great group are associated with the arctic tundra biome?

_____

_____

Exercise 10-A Soil Textures

[Text p. 343–344, Figures 10.2, 10.3.]

Did you ever wonder what makes those "Idaho" baking potatoes so wonderfully smooth and uniformly shaped? Careful plant breeding by agronomists, perhaps, or do the growers just select the best ones for market? Could the secret lie in a remarkable kind of soil in which they are grown? Wheat and potatoes are important agricultural exports of the Palouse region that lies in a part of the Columbia Plateau Province where Washington, Oregon, and Idaho meet.

Looking up the soil order and suborder of the Palouse region on the soil map of the United States (see Exercise 10-B, Map A), we learn that these soils are Xerolls, the suborder of Mollisols that is characteristic of the Mediterranean type of climate. These are naturally fertile soils with a loose texture, and that in itself could at least partly explain the fine potatoes. Xerolls, however, cover a much larger area in that part of the U.S. than just the potato region. Further research uncovers descriptions of two local soils: Palouse Soils and Walla Walla Soils. The U.S. Department of Agriculture text states that the parent matter of these soils is "largely loess (floury wind-blown material," and that "silt loam is the predominant texture."

Probing further, we consult a map of the aeolian deposits (dune sand and loess) of the United States. Over the Palouse region there is indeed a layer of loess; it covers 100 percent of the surface. It is Wisconsinan in age and ranges in thickness from 4 to 16 feet. Typical American loess of that age consists of about 90 percent silt and about 5 percent each of sand (mostly very fine sand) and clay. On the uplands, this silt fell directly from the turbid air of dust storms, while torrential winter rains washed some of it into the valley bottoms, making the Walla Walla Soils. Here, where irrigation is feasible during the dry, hot summers, the remarkable potatoes are grown. The tubers are free to grow to perfection in the soft silt that surrounds them. And don't forget the huge, sweet Walla Walla onions!

Other soil textures can be unfavorable to agriculture, and even make the soil next-to-impossible to cultivate. Vertisols, rich in clays that swell and become sticky upon being soaked with water, also harden, shrink, and crack upon drying out. Over the centuries, vertisols of Africa and India have resisted the primitive plows drawn by humans and animals, yet their natural soil fertility is extremely high.

Modern soil science is rigorous in its precise measurements of the physical and chemical properties of soils. Our exercise illustrates how this quantitative approach is applied to the textures of parent materials of the soil.

Soil Texture Grades

Soil texture grades as defined by the U.S. Department of Agriculture are given in Table A. This information agrees with that of text Figure 10.2, but contains additional subdivisions of the sand and gravel grades.

Table A *Soil Texture Grades*

Name of grade	Diameter (mm)
Coarse gravel	Above 2.0
Fine gravel	1.0–2.0
Coarse sand	0.5–1.0
Medium sand	0.25–0.5
Fine sand	0.1–0.25
Very fine sand	0.05–0.1
Silt	0.002–0.05
Clay	Below 0.002 (2.0 microns)
Colloidal Clay	Below 0.0001 (0.01 micron.

[Data source: U. S. Department of Agriculture.]

Soil Texture Classes

Soil texture classes make use of the scale of texture grades, combining the percentages of three grades: sand, silt, clay. Text Figure 10.3 shows the components of five of the classes, using pie-diagrams. We reproduce this illustration here as Figure A.

Figure B enlarges on this concept to cover all texture classes used by the USDA. It is a *triangular graph* that simultaneously accommodates three sets of numbers. All possible combinations of three percentages, ranging from 0 to 100 can be plotted as points. We have plotted the five examples of Figure A as labeled points. Check each of these points to see that the percentages are correctly plotted.

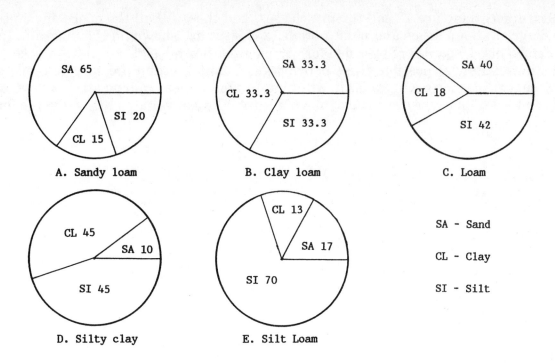

Figure A Pie-diagrams of five examples of soil texture classes.

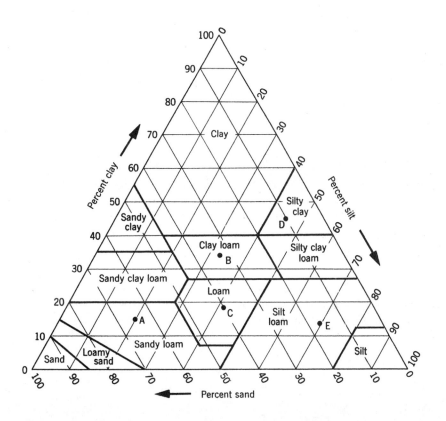

Figure B Soil texture diagram. (From U.S. Department of Agriculture, Soil Survey Staff, 1975, **Soil Taxonomy,** Handbook No. 436, p. 471, Figure 38.)

(1) Altogether, there are 12 soil texture classes, and these fill all the space on the triangular diagram. Boundaries and names of all 12 classes are shown on the diagram. In the table below, blank spaces are provided for percentages of sand, silt, and clay for each class. Estimate as closely as possible these percentages from the graph. Be sure you follow the correct set of lines for each grade. It will help to hold a clear plastic straightedge along the correct set of lines for each grade, moving it from lowest value to highest value for the class.

Class name	Sand (%)	Silt (%)	Clay (%)
Sand	_____	_____	_____
Loamy sand	_____	_____	_____
Sandy loam	_____	_____	_____
Sandy clay loam	_____	_____	_____
Loam (example)	43–52	28–50	7–27
Silt loam	_____	_____	_____
Silt	_____	_____	_____
Silty clay loam	_____	_____	_____
Clay loam	_____	_____	_____
Sandy clay	_____	_____	_____
Silty clay	_____	_____	_____
Clay	_____	_____	_____

(2) Using the data from the pie-diagrams in Figure A, plot the percentages of each texture grade on Graph A. Represent the percentage by a horizontal line drawn across the zone covered by the grade. If possible, use a different color for each sample. Label the lines with the corresponding letters A, B, C, etc.

(3) Table B gives the percentages of sand, silt, and clay in each of five soil samples. Locate each sample on the triangular diagram, Figure B. Make a dot, surrounded by a small circle, and label with the sample number. In the spaces below, write the name of the soil texture class.

Sample 1 _____   Sample 4 _____

Sample 2 _____   Sample 5 _____

Sample 3 _____

Table B Texture Classes of Five Samples

Sample No.	Sand (%) (arc)	Silt (%) (arc)	Clay (%) (arc)
1	15 54°	51 184°	34 122°
2	72 260°	14 50°	14 50°
3	10 36°	85 306°	5 18°
4	18 65°	32 115°	50 180°
5	47 169°	32 115°	21 76°

(4) Using the blank circles below, construct a pie-diagram of each of the above five samples. Note that beside each percentage in Table B is the arc, in degrees, for each sector in the pie. Use a protractor. Start at the horizontal zero reference radius in the 3-o'clock position, shown on each circle. The sector for sand is laid off anticlockwise from that reference line; the silt sector is laid off clockwise from that line. The clay sector will automatically fall between the sand and silt sectors, but you should measure it with the protractor as a check against error.

(5) Refer back to the third paragraph of this exercise, where the composition of a typical American loess soil is given. Find and label this point on the triangular diagram, Figure B. Name the texture class to which it belongs.

Texture class: _____

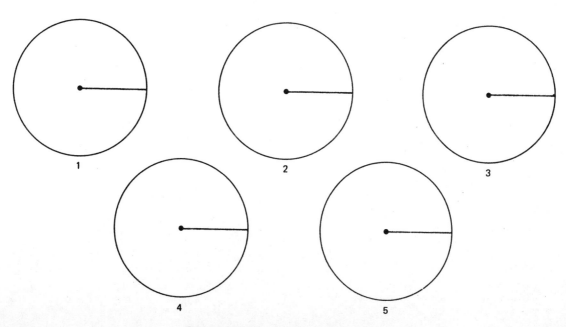

NAME _____   DATE _____

Exercise 10-B Soil Orders and Climate

[Text p. 302–365, Table 10.2, Figures 10.13, 10.14, 10.15, 10.16.]

Physical geography brings together its several subject areas in meaningful ways. Your study of world climates didn't just end with the last page of text Chapter 10. Now we reach back to world climates for relationships between soil orders and climate types. Climates will be referred to many times more in the chapters on landforms, because climate influences the processes that shape landforms.

We begin by relating the soil profiles of text Figure 10.14 to the climate type of the place from which each was taken. For this purpose, we offer three special maps, none of which is in your textbook. Maps A and B show the United States and the adjacent part of Canada lying south of the 51st parallel. Map A shows soil orders and some of the most important suborders. Map B shows climates of the same area, using the same boundaries and climate types as in text Chapter 7 and Figure 7.6. These two maps will enable you to locate places much more closely in terms of state and province boundaries, and other reference features such as rivers and lakes. Map C is a detailed world soils map showing suborders and their various combinations.

(1) Taking the soil profile photographs, text Figure 10.14, in order from A through K, find the location of each on the U.S.-Canada soils map, Map A, and enter the identifying profile letter on the map. Be sure the letter falls within the map area of the correct soil order or suborder. For foreign countries, use the world soils map, Map C. Find each locality on the U.S.-Canada climate map, Map B, or the world climate map, text Figure 7.6. Enter the climate numbers in the blank spaces below.

Note: You are not required to know the suborders marked by an asterisk, as they are not shown on your textbook soils map, Figure 10.13, and they are not explained in your textbook.

Profile	Soil Order	Soil Suborder	Climate Number	Type & Subtype Name
A	Oxisols	Torrox*	____	_____
B	Ultisols	Udult*	____	_____
C	Vertisols	Ustert*	____	_____
D	Alfisols	Udalf	____	_____
E	Alfisols	Ustalf	____	_____

(Continued on following page.)

F	Spodosols	Orthod*	___	_____
G	Mollisols	Boroll	___	_____
H	Mollisols	Udoll	___	_____
I	Mollisols	Ustoll	___	_____
J	Mollisols	Rendoll*	___	_____
K	Aridisols	Argid*	___	_____

Oxisols

(2) Compare the Oxisol profile of text Figure 10.14A with that in Figure 10.15. In what way are the two soil profiles similar?

_____

_____

_____

_____

(3) On the world soils map, text Figure 10.13, determine the latitudinal range of the Oxisols. What is their northern limit; their southern limit? Name the locality in which each limit lies.

Northern limit: _____ Place: _____

Southern limit: _____ Place: _____

Utisols

(4) Compare the Oxisol profile (A) with that of the Ultisol (B). What is the most important difference between the two? Before answering, examine also text Figure 10.16 and read the figure legend.

_____

_____

_____

_____

Copyright © by Arthur N. Strahler

(5) Find and describe the plinthite zone in text Figures 10.14B and 10.16. At what depth below the surface does the plinthite first appear?

_____

_____

_____

(6) On the world soils map, text Figure 10.13, determine the northern and southern latitudinal limits of the Ultisols. Compare your findings with those for the latitude limits of the Oxisols.

Northern limit: _____   Place: _____

Southern limit: _____   Place: _____

_____

_____

_____

(7) Following up on Question 6, it is obvious from the world soils map that by far the largest area of Ultisols is in low latitudes, i.e., in the equatorial, tropical, and subtropical zones. Using the world climate map, text Figure 7.6, list below the low-latitude climates in which important areas of Ultisols are found.

_____

_____

Vertisols

(8) Study the surface details of the Vertisol profile in text Figures 10.14C. In your opinion, are horizons clearly visible as horizontal layers? What minor features of soil structure are clearly shown in these profiles? What is the origin of these structures?

_____

_____

_____

_____

(9) The accompanying schematic drawing, Figure A, shows how Vertisols evolve by the alternate widening and closing of deep soil cracks. During this annual cycle of change, the soil "swallows itself." In the proper space below the drawing, label the following: Dry season; Wet season. Three stages are represented by three cracks, each undergoing a different process. Below each of the three, enter a description of what is taking place. (Make three columns.)

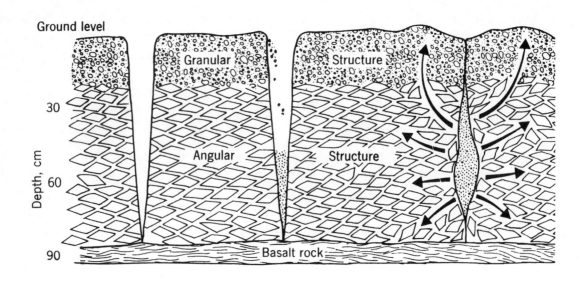

Stage A   Stage B   Stage C

_____   _____

_____   _____   _____

_____   _____   _____

_____   _____   _____

_____   _____   _____

Figure A *Schematic diagram of the effects of seasonal shrinking and swelling of the soil to produce the structure of the Vertisols. (From H. D. Foth,* Fundamentals of Soil Science, *p. 281, Figure 10-23. Copyright © 1984 by John Wiley & Sons, New York. Reprinted by permission of the author and John Wiley & Sons, Inc.)*

Alfisols

As explained in your textbook (p. 355–357), the Alfisols are subdivided into four suborders, each representing a different soil-water regime: Boralfs, Udalfs, Ustalfs, and Xeralfs. The world soils map, text Figure 10.13, shows the distributions of these suborders. Map C also shows the suborders.

(10) Boralfs are not shown in photos in your textbook, but they are a major suborder in terms of extent. Find the areas of Boralfs on the world soils map, text Figure 10.13. Describe the areas in terms of latitude and continental location. What soil order lies adjacent to the Boralfs on the low-latitude side?

_____

_____

_____

_____

(11) Compare the areas of Boralfs with the climate map, Figure 7.6. With what climate types are they closely associated? (Give number/letter codes.) Are these rated as moist or dry climates?

_____

_____

_____

(12) Uldalfs, the moist climate suborder of Alfisols, are represented by text Figure 10.14D, an example from Michigan. What important horizons does this profile display?

_____

_____

_____

(13) With what two climate types are the Udalfs most closely associated in middle latitudes? (Give codes.)

_____

_____

(14) Study the distribution of the Ustalfs as shown on the world soils map, text Figure 10.13. (a) In what latitude zones do the large areas lie? (b) Name the countries or continental regions in which these large areas occur (from east to west, from northern to southern hemisphere). (c) Name the climate types found in these regions.

(a) Latitude zones: _____

(b) Countries or continental regions: _____

_____

_____

(c) Climate types: _____

_____

(15) Xeralfs are Alfisols, found in the regions of Mediterranean climate (7). Find these regions on the world soils map, Figure 10.13. List them by name as you did for the Ustalfs.

_____

_____

_____

Spodosols and Histosols

Spodosols are of wide extent in the northern hemisphere. Here, Histosols are found as isolated patches of bog or muck within the Spodosol areas. The two orders are commonly combined as one unit on the soils map. Spodosols and Boralfs also occur as a mixture of patches within the same area.

(16) Identify by geographical names the major areas of Spodosols shown on the world soils map, text Figure 10.13. Give the approximate latitudinal range of each.

_____

_____

_____

(17) With what climate type or types are these Spodosol regions associated?

_____

Mollisols

As explained in your textbook, p. 363, the Mollisols include four important suborders: Borolls, Udolls, Ustolls, and Xerolls. Profiles of the first three are shown in Figure 10.14. The Rendolls, a fifth suborder, are of much less importance on a world scale, they occur locally on limestones or other forms of carbonate parent matter. The profile example, Figure 10.14J, is from Argentina. We will concentrate on the four widespread orders. They are shown as separate areas on Maps A and C. We will limit our investigation of the Mollisols to North America. (We suggest that you color the maps areas of each suborder, using for each a different color of your own choice.)

(18) (a) Where in North America are the Borolls found? Name states and provinces. Give the latitudinal range. (b) With what climate types and subtypes are the Borolls associated? (Use Map B.)

(a) _____

_____

_____

(b) _____

_____

_____

(19) For the Udolls, give same information as for Question 18.

(a) _____

_____

_____

(b) _____

_____

(20 For the Ustolls, give the same information as for Question 18.

(a) _____

_____

_____

(b) _____

(21) For the Xerolls, give the same information as for Question 18.

(a) _____

_____

_____

(b) _____

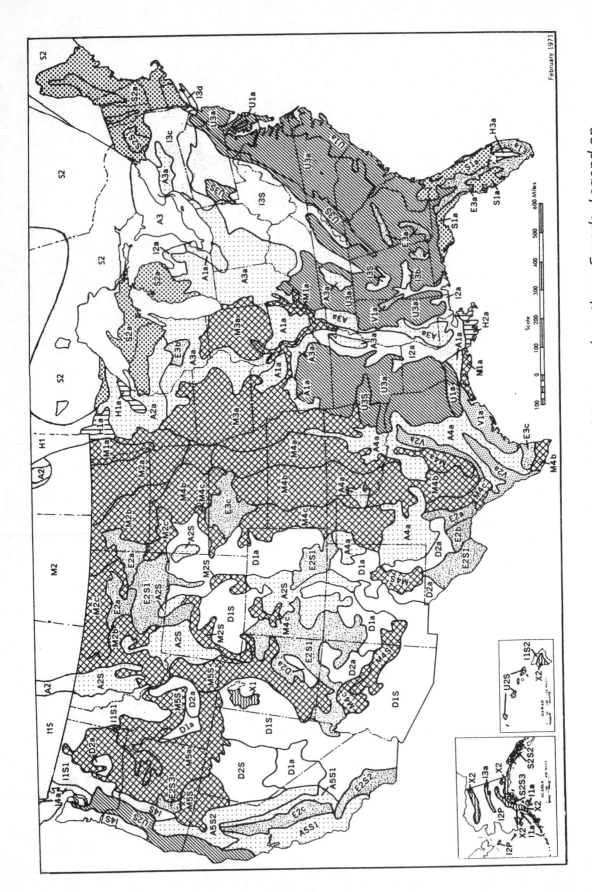

Map A *Soil orders and suborders of the United States and southern Canada. Legend on facing page. (Soil Conservation Service, U.S. Department of Agriculture.)*

ALFISOLS

AQUALFS
A1a—Aqualfs with Udalfs, Haplaquepts, Udolls; gently sloping.

BORALFS
A2a—Boralfs with Udipsamments and Histosols; gently and moderately sloping.
A2S—Cryoboralfs with Borolls, Cryochrepts, Cryorthods, and rock outcrops; steep.

UDALFS
A3a—Udalfs with Aqualfs, Aquolls, Rendolls, Udolls, and Udults; gently or moderately sloping.

USTALFS
A4a—Ustalfs with Ustochrepts, Ustolls, Usterts, Ustipsamments, and Ustorthents; gently or moderately sloping.

XERALFS
A5S1—Xeralfs with Xerolls, Xerorthents, and Xererts; moderately sloping to steep.
A5S2—Ultic and lithic subgroups of Haploxeralfs with Andepts, Xerults, Xerolls, and Xerochrepts; steep.

ARIDISOLS

ARGIDS
D1a—Argids with Orthids, Orthents, Psamments, and Ustolls; gently and moderately sloping.
D1S—Argids with Orthids, gently sloping; and Torriorthents, gently sloping to steep.

ORTHIDS
D2a—Orthids with Argids, Orthents, and Xerolls; gently or moderately sloping.
D2S—Orthids, gently sloping to steep, with Argids, gently sloping; lithic subgroups of Torriorthents and Xerorthents, both steep.

ENTISOLS

AQUENTS
E1a—Aquents with Quartzipsamments, Aquepts, Aquolls, and Aquods; gently sloping.

ORTHENTS
E2a—Torriorthents, steep, with borollic subgroups of Aridisols; Usterts and aridic and vertic subgroups of Borolls; gently or moderately sloping.
E2b—Torriorthents with Torrerts; gently or moderately sloping.
E2c—Xerorthents with Xeralfs, Orthids, and Argids; gently sloping.
E2S1—Torriorthents; steep, and Argids, Torrifluvents, Ustolls, and Borolls; gently sloping.
E2S2—Xerorthents with Xeralfs and Xerolls; steep.
E2S3—Cryorthents with Cryosamments and Cryandepts; gently sloping to steep.

PSAMMENTS
E3a—Quartzipsamments with Aquults and Udults; gently or moderately sloping.
E3b—Udipsamments with Aquolls and Udalfs; gently or moderately sloping.
E3c—Ustipsamments with Ustalfs and Aquolls; gently or moderately sloping.

HISTOSOLS

HISTOSOLS
H1a—Hemists with Psammaquents and Udipsamments; gently sloping.
H2a—Hemists and Saprists with Fluvaquents and Haplaquepts; gently sloping.
H3a—Fibrists, Hemists, and Saprists with Psammaquents; gently sloping.

INCEPTISOLS

ANDEPTS
I1a—Cryandepts with Cryaquepts, Histosols, and rock land; gently or moderately sloping.
I1S1—Cryandepts with Cryochrepts, Cryumbrepts, and Cryorthods; steep.
I1S2—Andepts with Tropepts, Ustolls, and Tropofolists; moderately sloping to steep.

AQUEPTS
I2a—Haplaquepts with Aqualfs, Aquolls, Udalfs, and Fluvaquents; gently sloping.
I2P—Cryaquepts with cryic great groups of Orthents, Histosols, and Ochrepts; gently sloping to steep.

OCHREPTS
I3a—Cryochrepts with cryic great groups of Aquepts, Histosols, and Orthods; gently or moderately sloping.
I3b—Eutrochrepts with Uderts; gently sloping.
I3c—Fragiochrepts with Fragioquepts, gently or moderately sloping; and Dystrochrepts, steep.
I3d—Dystrochrepts with Udipsamments and Haplorthods; gently sloping.
I3S—Dystrochrepts, steep, with Udalfs and Udults; gently or moderately sloping.

UMBREPTS
I4a—Haplumbrepts with Aquepts and Orthods; gently or moderately sloping.
I4S—Haplumbrepts and Orthods; steep, with Xerolls and Andepts; gently sloping.

MOLLISOLS

AQUOLLS
M1a—Aquolls with Udalfs, Fluvents, Udipsamments, Ustipsamments, Aquepts, Eutrochrepts, and Borolls; gently sloping.

BOROLLS
M2a—Udic subgroups of Borolls with Aquolls and Ustorthents; gently sloping.
M2b—Typic subgroups of Borolls with Ustipsamments, Ustorthents, and Boralfs; gently sloping.
M2c—Aridic subgroups of Borolls with Borollic subgroups of Argids and Orthids, and Torriorthents; gently sloping.
M2S—Borolls with Boralfs, Argids, Torriorthents, and Ustolls; moderately sloping or steep.

UDOLLS
M3a—Udolls, with Aquolls, Udalfs, Aqualfs, Fluvents, Psamments, Ustorthents, Aquepts, and Albolls; gently or moderately sloping.

USTOLLS
M4a—Udic subgroups of Ustolls with Orthents, Ustochrepts, Usterts, Aquents, Fluvents, and Udolls; gently or moderately sloping.
M4b—Typic subgroups of Ustolls with Ustalfs, Ustipsamments, Ustorthents, Ustochrepts, Aquolls, and Usterts; gently or moderately sloping.
M4c—Aridic subgroups of Ustolls with Ustalfs, Orthids, Ustipsamments, Ustorthents, Ustochrepts, Torriorthents, Borolls, Ustolls, and Usterts; gently or moderately sloping.
M4S—Ustolls with Argids and Torriorthents; moderately sloping or steep.

XEROLLS
M5a—Xerolls with Argids, Orthids, Fluvents, Cryoboralfs, Cryoborolls, and Xerorthents; gently or moderately sloping.
M5S—Xerolls with Cryoboralfs, Xeralfs, Xerorthents, and Xererts; moderately sloping or steep.

SPODOSOLS

AQUODS
S1a—Aquods with Psammaquents, Aquolls, Humods, and Aquults; gently sloping.

ORTHODS
S2a—Orthods with Boralfs, Aquents, Orthents, Psamments, Histosols, Aquepts, Fragiochrepts, and Dystrochrepts; gently or moderately sloping.
S2S1—Orthods with Histosols, Aquents, and Aquepts; moderately sloping or steep.
S2S2—Cryorthods with Histosols; moderately sloping or steep.
S2S3—Cryorthods with Histosols, Andepts and Aquepts; gently sloping to steep.

ULTISOLS

AQUULTS
U1a—Aquults with Aquents, Histosols, Quartzipsamments, and Udults; gently sloping.

HUMULTS
U2S—Humults with Andepts, Tropepts, Xerolls, Ustolls, Orthox, Torrox, and rock land; gently sloping to steep.

UDULTS
U3a—Udults with Udalfs, Fluvents, Aquents, Quartzipsamments, Aquepts, Dystrochrepts, and Aquults; gently or moderately sloping.
U3S—Udults with Dystrochrepts; moderately sloping or steep.

VERTISOLS

UDERTS
V1a—Uderts with Aqualfs, Eutrochrepts, Aquolls, and Ustolls; gently sloping.

USTERTS
V2a—Usterts with Aqualfs, Orthids, Udifluvents, Aquolls, Ustolls, and Torrerts; gently sloping.

Areas with little soil
X1—Salt flats.
X2—Rock land (plus permanent snow fields and glaciers).

Slope classes
Gently sloping—Slopes mainly less than 10 percent, including nearly level.
Moderately sloping—Slopes mainly between 10 and 25 percent.
Steep—Slopes mainly steeper than 25 percent.

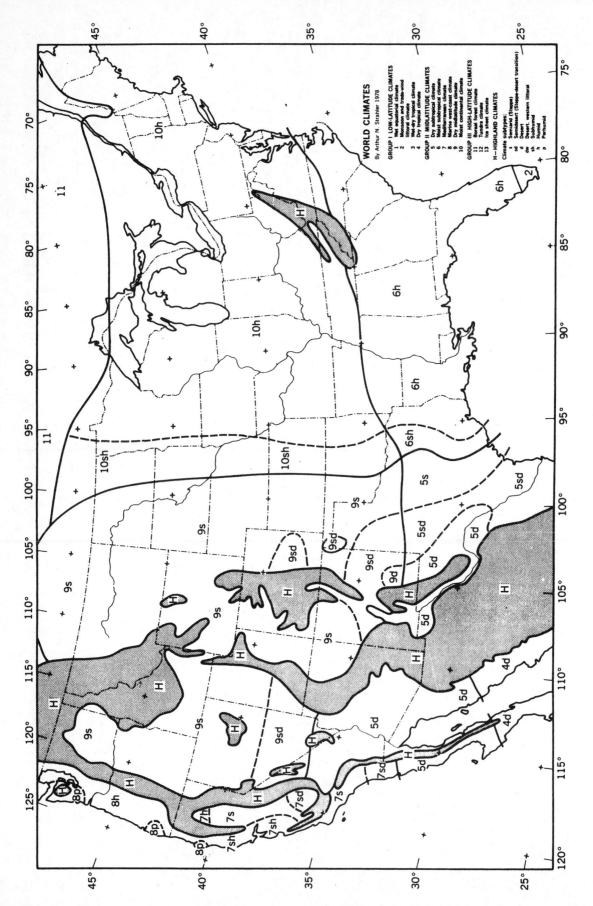

Map B Climate types and subtypes of the contiguous 48 United States and southern Canada. (A.N. Strahler.)

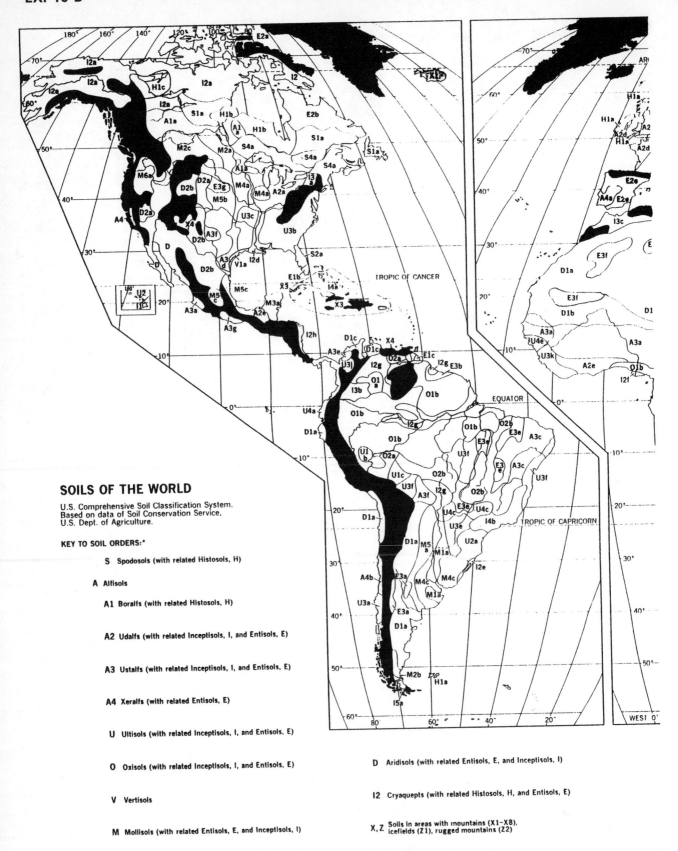

SOILS OF THE WORLD

U.S. Comprehensive Soil Classification System.
Based on data of Soil Conservation Service,
U.S. Dept. of Agriculture.

KEY TO SOIL ORDERS:*

S Spodosols (with related Histosols, H)

A Alfisols

  A1 Boralfs (with related Histosols, H)

  A2 Udalfs (with related Inceptisols, I, and Entisols, E)

  A3 Ustalfs (with related Inceptisols, I, and Entisols, E)

  A4 Xeralfs (with related Entisols, E)

  U Ultisols (with related Inceptisols, I, and Entisols, E)

  O Oxisols (with related Inceptisols, I, and Entisols, E)

  V Vertisols

  M Mollisols (with related Entisols, E, and Inceptisols, I)

D Aridisols (with related Entisols, E, and Inceptisols, I)

I2 Cryaquepts (with related Histosols, H, and Entisols, E)

X, Z Soils in areas with mountains (X1–X8),
icefields (Z1), rugged mountains (Z2)

***Map C** Soils of the World. From A.H. and A.N. Strahler, **Modern Physical Geography,**
4th Ed., 1992. Copyright © by John Wiley & Sons, Inc.*

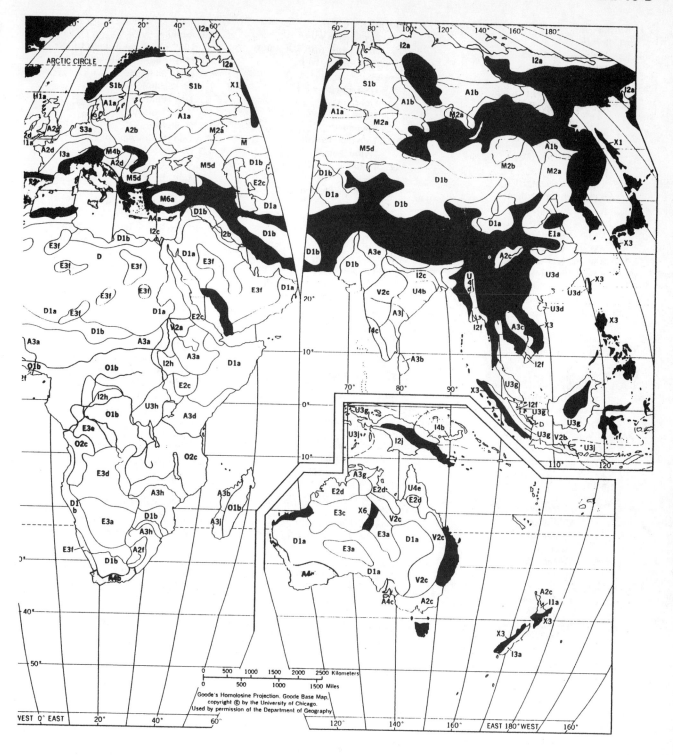

Goode's Homolosine Projection. Goode Base Map.
copyright © by the University of Chicago.
Used by permission of the Department of Geography

EX. 10-B

A   Alfisols
  A1   Boralfs
    A1a   with Histosols
    A1b   with Spodosols
  A2   Udalfs
    A2a   with Aqualfs
    A2b   with Aquolls
    A2c   with Hapludults
    A2d   with Ochrepts
    A2e   with Troporthents
    A2f   with Udorthents
  A3   Ustalfs
    A3a   with Tropepts
    A3b   with Troporthents
    A3c   with Tropustults
    A3d   with Usterts
    A3e   with Ustochrepts
    A3f   with Ustolls
    A3g   with Ustorthents
    A3h   with Ustox
    A3j   Plinthustalfs with
          Ustorthents
  A4   Xeralfs
    A4a   with Xerochrepts
    A4b   with Xerorthents
    A4c   with Xerults

D   Aridisols
  D1   Aridisols, undifferentiated
    D1a   with Orthents
    D1b   with Psamments
    D1c   with Ustalfs
  D2   Argids
    D2a   with Fluvents
    D2b   with Torriorthents

E   Entisols
  E1   Aquents
    E1a   Haplaquents with
          Udifluvents
    E1b   Psammaquents with
          Haplaquents
    E1c   Tropaquents with
          Hydraquents
  E2   Orthents
    E2a   Cryorthents
    E2b   Cryorthents with Orthods
    E2c   Torriorthents with
          Aridisols
    E2d   Torriorthents with Ustalfs
    E2e   Xerorthents with Xeralfs
  E3   Psamments
    E3a   with Aridisols
    E3b   with Orthox
    E3c   with Torriorthents
    E3d   with Ustalfs
    E3e   with Ustox
    E3f   of shifting sands
    E3g   Ustipsamments with
          Ustolls

H   Histosols
  H1   Histosols, undifferentiated
    H1a   with Aquods
    H1b   with Boralfs
    H1c   with Cryaquepts

I   Inceptisols
  I1   Andepts
    I1a   Dystrandepts with
          Ochrepts

I2   Aquepts
    I2a   Cryaquepts with Orthents
    I2b   Halaquepts with Salorthids
    I2c   Haplaquepts with
          Humaquepts
    I2d   Haplaquepts with
          Ochraqualfs
    I2e   Humaquepts with
          Psamments
    I2f   Tropaquents with
          Hydraquents
    I2g   Tropaquepts with
          Plinthaquults
    I2h   Tropaquepts with
          Tropaquents
    I2j   Tropaquepts with
          Tropudults

I3   Ochrepts
    I3a   Dystrochrepts with
          Fragiochrepts
    I3b   Dystrochrepts with Orthox
    I3c   Xerochrepts with Xerolls

I4   Tropepts
    I4a   with Ustalfs
    I4b   with Tropudults
    I4c   with Ustox

I5   Umbrepts
    I5a   with Aqualfs

M   Mollisols
  M1   Albolls
    M1a   with Aquepts
  M2   Borolls
    M2a   with Aquolls
    M2b   with Orthids
    M2c   with Torriorthents
  M3   Rendolls
    M3a   with Usterts
  M4   Udolls
    M4a   with Aquolls
    M4b   with Eutrochrepts
    M4c   with Humaquepts
  M5   Ustolls
    M5a   with Argialbolls
    M5b   with Ustalfs
    M5c   with Usterts
    M5d   with Ustochrepts
  M6   Xerolls
    M6a   with Xerorthents

O   Oxisols
  O1   Orthox
    O1a   with Plinthaquults
    O1b   with Tropudults
  O2   Ustox
    O2a   with Plinthaquults
    O2b   with Tropustults
    O2c   with Ustalfs

S   Spodosols
  S1   Spodosols, undifferentiated
    S1a   cryic regimes, with Boralfs
    S1b   cryic regimes, with
          Histosols
  S2   Aquods
    S2a   Haplaquods with
          Quartzipsamments
  S3   Humods
    S3a   with Hapludalfs
  S4   Orthods
    S4a   Haplorthods with Boralfs

U   Ultisols
  U1   Aquults
    U1a   Ochraquults with Udults
    U1b   Plinthaquults with Orthox
    U1c   Plinthaquults with
          Plinthaquox
    U1d   Plinthaquults with
          Tropaquepts
  U2   Humults
    U2a   with Umbrepts
  U3   Udults
    U3a   with Andepts
    U3b   with Dystrochrepts
    U3c   with Udalfs
    U3d   Hapludults with
          Dystrochrepts
    U3e   Rhodudults with Udalfs
    U3f   Tropudults with Aquults
    U3g   Tropudults with
          Hydraquents
    U3h   Tropudults with Orthox
    U3j   Tropudults with Tropepts
    U3k   Tropudults with Tropudalfs
  U4   Ustults
    U4a   with Ustochrepts
    U4b   Plinthustults with
          Ustorthents
    U4c   Rhodustults with Ustalfs
    U4d   Tropustults with
          Tropaquepts
    U4e   Tropustults with Ustalfs

V   Vertisols
  V1   Uderts
    V1a   with Usterts
  V2   Usterts
    V2a   with Tropaquepts
    V2b   Tropofluvents
    V2c   with Ustalfs

X   Soils in areas with mountains
  X1 Cryic great groups of Entisols,
       Inceptisols, and Spodosols.
  X2 Boralfs and cryic great groups of
       Entisols and Inceptisols.
  X3 Udic great groups of Alfisols,
       Entisols, and Ultisols;
       Inceptisols.
  X4 Ustic great groups of Alfisols,
       Inceptisols, Mollisols, and
       Ultisols.
  X5 Xeric great groups of Alfisols,
       Entisols, Inceptisols, Mollisols,
       and Ultisols.
  X6 Torric great groups of Entisols;
       Aridisols.
  X7 Histic and cryic great groups of
       Alfisols, Entisols, Inceptisols, and
       Mollisols; ustic great groups of
       Ultisols; cryic great groups of
       Spodosols.
  X8 Aridisols; torric and cryic great
       groups of Entisols, and cryic
       great groups of Spodosols and
       Inceptisols.

Z   Miscellaneous
  Z1 Ice sheets
  Z2 Rugged mountains, mostly
       devoid of soil (includes glaciers,
       permanent snowfields, and in
       some places, areas of soil.)

166

Copyright © by Arthur N. Strahler

NAME _____   DATE _____

Exercise 11-A The Igneous Rocks and Their Minerals

[Text p. 374–379, Table 11.1, Figures 11.4, 11.5.]

Geologists keep looking for the oldest rocks in the earth's crust, and every few years a new world's record is claimed. The ancient shield of western Greenland and eastern Labrador has been one of the best hunting grounds. In a mountainous region called Isua, close to the edge of the Greenland Ice Sheet, are some strongly deformed metamorphic rocks that were once ordinary extrusive volcanic and sedimentary rocks. Their age: a record 3.8 billion years! But they aren't the oldest rocks of our planet, because the marine-type sediments in them must have been derived from even older rock.

Individual grains of the mineral zircon within some ancient Australian metamorphic rocks (quartzites) have yielded ages between 4.1 and 4.2 billion years. Zircon, a silicate of the element zirconium, is present as a secondary mineral in most igneous rocks, and especially common in the granitic (felsic) group. Zircon is also extremely durable, and grains of the mineral can travel long distances in sand and gravel of streams. That's a good reason to infer that the first crustal rocks were igneous types, among them granite, intruded over 4 billion years ago.

The Granite-Gabbro Series Information in text Figure 11.4 provides good basic material on igneous rocks and their silicate minerals. The information in that diagram is descriptive, using only words and pictures. We will now go a step further and make this information quantitative. How much of each of the minerals shown is found in each of the important igneous rock varieties? This information is shown in Figure A. Study it carefully to understand how it is constructed.

(1) What is the meaning of the horizontal lines drawn across the graph in Figure A? What units are used to express the quantity of each important mineral in the rock?

_____

_____

_____

_____

_____

EX. 11-A

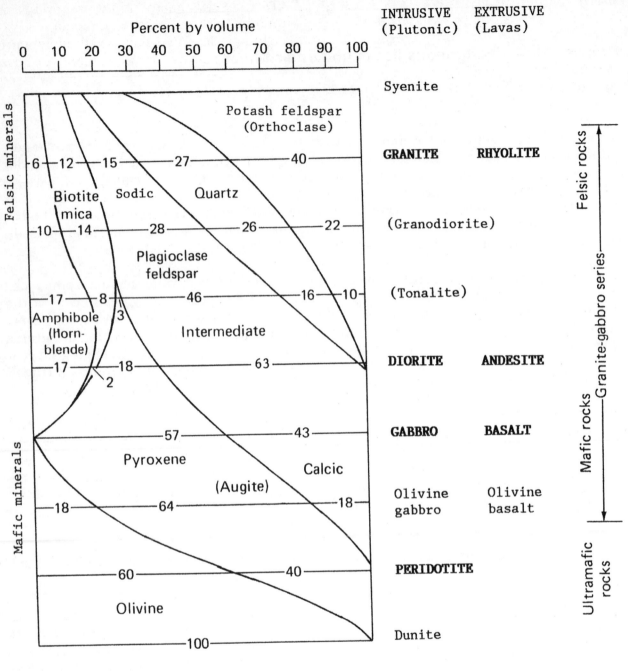

Figure A

(2) Explain the curved lines that cut across the horizontal lines on the graph.

More About the Feldspars Text Figure 11.4 describes two major kinds or groups of feldspars: One is potash feldspar, a silicate of aluminum rich in potassium. A common variety of potash feldspar is orthoclase. The other is plagioclase feldspar, which is actually a group of several distinctly different minerals. All are silicates of aluminum; all have some of both calcium and sodium in their composition. To understand our graph, we need to elaborate on the plagioclase group. It consists of three kinds or classes:

Sodic plagioclase: rich in sodium; poor in calcium. A common variety is <u>albite</u>.

Intermediate plagioclase: about equal parts of calcium and sodium. A common variety is *labradorite*.

Calcic plagioclase: rich in calcium; poor in sodium. A common variety is *anorthite*.

We will find that each of these three groups has a different association with the igneous rock varieties.

(3) A bracket at the right of the graph (Figure A) designates the granite-gabbro series. Name the intrusive igneous rocks in this series in order from top to bottom and name the two most abundant minerals in each. Give percentages of each mineral. Follow with the ultramafic rocks. Include only those rocks in bold capitals.

Name	Mineral 1	Mineral 2
_____	_____	_____
_____	_____	_____
_____	_____	_____
_____	_____	_____

Five other igneous rocks are named on the graph for the sake of completeness, but you may ignore them so as to concentrate on the four major kinds described in your textbook. Knowledge of the four major kinds and their extrusive equivalents (lavas) will be put to good use in Chapter 12 to explain the kinds of crust and their origin through processes of plate tectonics.

Rock Densities Perhaps the most important message you can read from Figure A is the typical density of each of the four major igneous rock types. The rock density can be estimated from the mineral percentages given in the graph. Each percentage must be multiplied by its corresponding mineral density and the average density calculated as in the following example for granite. Use density values given in Table A.

(4) Complete the following table to find the density of each igneous rock listed.

	Granite	Diorite	Gabbro	Peridotite
Quartz	0.27 x 2.65 = 0.716			
Potash feldspar	0.40 x 2.57 = 1.028			
Plagioclase feldspar	0.15 x 2.62 = 0.393			
Biotite mica	0.12 x 3.00 = 0.360			
Amphibole (hornbende)	0.06 x 3.20 = 0.192			
Pyroxene (augite)	(none)			
Olivine	(none)			
Rock density:	2.689			
(rounded off)	2.69			

Table A Mineral densities

Quartz	2.65 (g/cc)
Potash feldspar (orthoclase)	2.57
Plagioclase feldspars:	
Sodic (albite)	2.62
Intermediate (labradorite)	2.71
Calcic (anorthite)	2.76
Biotite mica (average)	3.00
Amphibole (hornblende)	3.20
Pyroxene (augite)	3.30
Olivine (common)	3.40

(5) Offer a general statement of the relationship between mineral varieties present in the rock and its density.

_____

_____

_____

_____

_____

Igneous Rocks under the Microscope Figure B is a drawing of the appearance of mineral grains of two different kinds of igneous rocks seen under the microscope. Enlargement is about 10 times actual size. Very thin slices of the rock are mounted on a slide and illuminated from below by polarized light. Notice that the patterns of the minerals resemble those shown in text Figure 11.4. You can identify the minerals in the slides by comparing texture patterns with those in Figure 11.4.

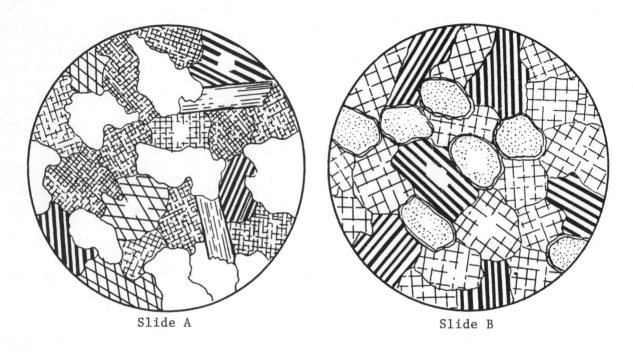

Slide A   Slide B

Figure B

(6) Using the code letters for seven minerals given below, list the minerals found in each slide. On the slide illustration write the code letter on at least one example of a grain of each mineral variety present.

Code	Mineral name	Slide A	Slide B
Q	Quartz	____	____
K	Potash feldspar	____	____
P	Plagioclase feldspar	____	____
B	Biotite mica	____	____
H	Hornblende (amphibole group)	____	____
A	Augite (pyroxene group)	____	____
O	Olivine	____	____

(7) Identify the varieties of igneous rock shown in the slides. Give the names below:

Slide A: _____   Slide B: _____

Exercise 11-B Size Grades of Sediment Particles

[Text p. 382–384.]

Mineral particles that make up clastic sedimentary rock were at one time in a loose state, spread in layers over continental surfaces in direct contact with the atmosphere or submerged under water on the floors of lakes, streams, coastal estuaries and lagoons, or oceans. Just how these particles are transported is a question we deal with at several points in later chapters. Running water, wind, waves and currents, and glacial ice are prime movers of sediment, each having its own unique form of action and leaving a distinctive form of sediment layers.

To prepare for what is to come in later chapters, we concentrate here on the vast range in sizes of the individual particles of sediment. Scientists not only describe discrete objects in words, but find it necessary to measure objects in terms of length, area, and volume. The study of sediments—sedimentology, that is—is a quantitative form of science. Classes of sizes are set up to make this task easier and more meaningful.

Particle Size and Surface Area To get acclimated to the world of particles, we start with a geometrical concept that is of great importance in both the geology of sediments and the science of agricultural soils. The mechanical disintegration of rock to form sediment (Chapter 14) is one of "making little ones out of big ones." As this process proceeds, an important effect is that the surface area of all the particles contained in a cube of given size, such as 1 cm on a side, increases very greatly as the particles are made smaller. Chemical alteration of minerals takes place on the exposed mineral surfaces, so the smaller the particles, the greater is the speed with which alteration can proceed.

Assume that we start with a single mineral cube 1 cm in height, width, and length. Imagine that we can slice this cube into perfect cubes of smaller and smaller dimensions in steps of powers of ten. What then will be (a) the increase in number of particles and (b) the increase in total surface area of all the particles?

(1) Fill in the missing figures in the following table:

Surface Area Resulting from the Subdivision of a One-Centimeter Cube

Cube dimension (length, cm)	Reciprocal of length	Particle name	Number of particles	Total surface area (cm^2)
1	1	Pebble	1	6
0.1	10	Coarse sand	10^3	60
0.01	_____	Silt	_____	_____
0.000,1	_____	Fine clay	_____	_____
0.000,01	_____	Colloidal clay	_____	_____

(2) Describe below the rate of increase in number of particles and in total surface area with respect to the cube dimension. (The answer requires knowledge of powers of ten.)

The number of particles increases as the _____ of the _____

of the cube dimension.

The total surface area is equal to _____ times the _____ of the cube

dimension.

Note: Colloidal particles of clay are chemically active because they bear electrical charges, by means of which they can hold and exchange chemical ions. This important property of colloids was emphasized in text Chapter 10 (p. 344, Figure 10.4).

The Wentworth Scale of Particle Grades Grades of mineral particles are most commonly described in terms of the Wentworth scale (named after a prominent geologist who did pioneering research on sediments). This scale is reproduced here as Table A, with the names of the grades, their sub-divisions, and the diameter in millimeters that marks the boundary between each grade and subgrade. Equivalents in inches are given for the coarser grades; in microns for the finer grades.

(3) Study the progression of diameters downward from boulders, through cobbles, pebbles, and sand. Describe this progression.

_____

_____

Table A *The Wentworth Scale of Size Grades*

Grade Name		mm	in.
		— 4096 —	— 160 —
	Very large		
		— 2048 —	— 80 —
	Large		
Boulders		— 1024 —	— 40 —
	Medium		
		— 512 —	— 20 —
	Small		
		— 256 —	— 10 —
	Large		
Cobbles		— 128 —	— 5 —
	Small		
		— 64 —	— 2.5 —
	Very coarse		
		— 32 —	— 1.3 —
	Coarse		
		— 16 —	— 0.6 —
Pebbles	Medium		
		— 8 —	— 0.3 —
	Fine		
		— 4 —	— 0.16 —
	Very fine		
		— 2 —	— 0.08 —
	Very coarse		Microns
		— 1 —	— 1000 —
	Coarse		
		— 0.5 —	— 500 —
Sand	Medium		
		— 0.25 —	— 250 —
	Fine		
		— 0.125 —	— 125 —
	Very fine		
		— 0.0625 —	— 62 —
	Coarse		
		— 0.0312 —	— 31 —
	Medium		
Silt		— 0.016 —	— 16 —
	Fine		
		— 0.008 —	— 8 —
	Very fine		
		— 0.004 —	— 4 —
	Coarse		
		— 0.00? —	— 2 —
	Medium		
Clay		— 0.001 —	— 1 —
	Fine		
		— 0.0005 —	— 0.5 —
	Very fine		
		— 0.00024 —	— 0.24 —
	(Colloids down to 0.001 microns)		

For the particle grades in the range from coarse sand down through silt, a natural sample containing a mixture of grades is separated out into grades by passing the sample through a series of wire sieves. The mesh openings of each sieve corresponds to one of the boundary values between grades. Particles larger than the sieve mesh diameter are caught in the sieve, while those that are smaller pass on through to the next sieve.

(4) Figure A shows the outlines of several mineral grains, enlarged ten times. A length scale is included. Measure the maximum diameter of each grain. In the spaces below, enter the grain diameter and the name of the grade, according to the Wentworth scale.

Code letter on grain	Max. diameter (mm)	Wentworth grade name
A	_____	_____
B	_____	_____
C	_____	_____
D	_____	_____
E	_____	_____
F	_____	_____
G	_____	_____
H	_____	_____
I	_____	_____
J	_____	_____

(5) Examine text Figure 10.2 (p. 344), showing size grades as defined by the U.S. Department of Agriculture for soil texture descriptions. Compare the size boundaries it uses with those of the Wentworth scale. Be specific.

_____

_____

_____

_____

_____

(6) Using horizontal lines, show the positions of the U.S.D.A. boundaries on the Wentworth scale diagram, Table A. Label the diameter values. Write the names of the USDA classes along the side of the diagram.

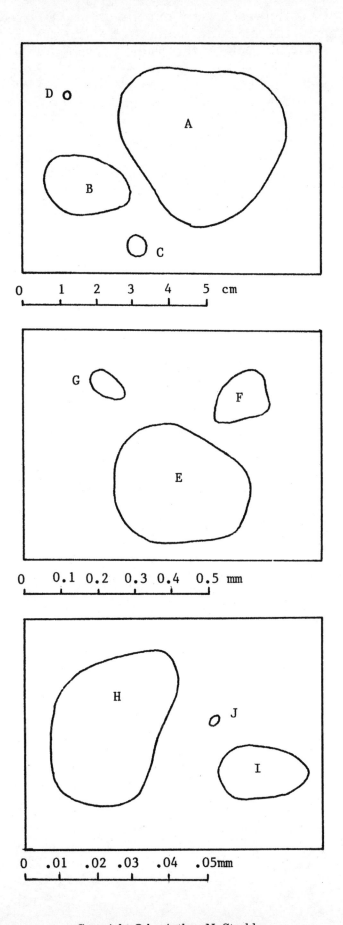

NAME _____   DATE _____

Exercise 11-C Geologic Maps and Structure Sections

[Text p. 376–378, Figure 11.10.]

A large portion of scientific research carried out by geologists consists of making special-ized maps of the earth's surface. These *geologic maps* show the areal distribution of dif-ferent kinds and/or ages of the bed rock that is exposed at the surface or lies beneath a thin cover of soil, weathered rock materials, or other shallow kinds of overburden. In a real sense, this map-making process is a geographical activity—showing the areal distri-bution of some property.

Geologic mapping requires (in most cases) direct examination of the surface rock, includ-ing tests performed directly on the rock, and often the taking of small rock samples for laboratory analysis. Modern techniques of remote sensing, ranging from the study of stereo-scopic photos to the analysis and interpretation of satellite images, provide valuable data for reconnaissance geologic maps or maps of special rock and mineral properties.

Traditional geology has required maps of two kinds of information: (a) the kind or variety of rock present (lithology), and (b) the age of the rock in terms of the name of the age (pe-riod, epoch, stage, etc.). In classical geology, fossils have been the basis for determining relative rock age. Today, radiometric methods can establish actual age in terms of years-before-present.

Figure A is a geologic map of a small area. Below it is a *structure section*, which is a side view of an imaginary vertical slice through the map along a selected line of traverse. In this case the, structure section coincides with the bottom line of the map. Structure sec-tions are partly inferred from the surface arrangements of the rock units, but are often supplemented by information from rock cores obtained by drilling into the crust. Seismic refraction investigation, using controlled seismic waves, is also very useful in revealing the rock structures far below the surface.

On our map, all three major classes of igneous rock can be found: Igneous, sedimentary, and metamorphic. The individual rock types within each type are listed on the map leg-end, but arranged in a random order.

Published geologic maps usually make use of colors instead of (or in addition to) line or dot patterns. We have suggested a suitable color for each rock type on the map. Coloring the map, cross section, and legend with color crayons or pencils is an optional step that may help you analyze the map patterns.

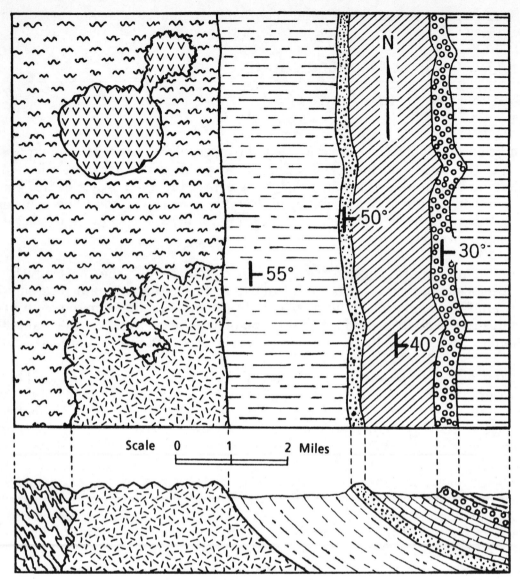

Scale 0   1   2 Miles

Vertical Structure Section

Rock units:
(In random order)

Limestone	Sandstone
Conglomerate	Lava	Gneiss
Grantite	Shale B	Shale A

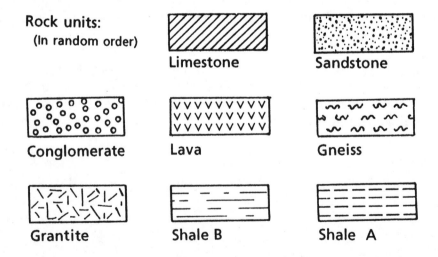

Figure A

(1) In the first blank space beside the rock name in the table below insert the class affiliation of each rock type, using the code: **IGN** = Igneous, **SED** = Sedimentary, **MET** = Metamorphic.

Rock name	Rock class	Order by Age	Color
Conglomerate	_____	_____	Orange
Grantite	_____	_____	Pink
Limestone	_____	_____	Blue
Lava	_____	_____	Red
Shale B	_____	_____	Green (deep)
Sandstone	_____	_____	Yellow
Gneiss	_____	_____	Violet
Shale A	_____	_____	Green (pale)

Your next task is to arrange the eight rock types in order of their geologic age. This requires that you use your knowledge of the origin of each rock type and combine this with the geometric relationship of each rock type with those that lie adjacent to it in the map and structure section.

(2) In the second set of blank spaces in the above table insert numbers 1 through 8 to indicate relative age of each type, the oldest being **1** and the youngest **8**. Hold off entering these numbers in final form until you have answered the following questions 3 through 6.

(3) Of all the rock types shown, select the very youngest. Enter the number **1** in the blank next to your choice. Explain below how you arrived at this decision.

_____

_____

_____

_____

(4) For the sedimentary rock types, what evidence and reasoning did you use to arrange them in relative order?

_____

_____

_____

_____

_____

_____

(5) Of two adjacent rock types—gneiss and granite—which one is the older? Give your reasons for reaching this conclusion.

_____

_____

_____

_____

_____

_____

(6) Of all the eight rock types, which did you pick out to be the oldest? Explain your choice.

_____

_____

_____

_____

_____

(7) The strata are shown as strongly inclined layers. Why are the edges of the layers abruptly ended at the ground surface? Explain. (Refer to text Figure 17.1.)

_____

_____

_____

_____

_____

(8) What is the meaning of the T-shaped map symbol and the numeral beside it? (Refer to text Figure 17.3.)

_____

_____

_____

_____

NAME _____ DATE _____

Exercise 12-B Kinds of Plate Junctions and Their Meaning

[Text p. 406–409, Figure 12.12, 12.13, 12.15.]

Lithospheric plates may seem to be highly varied in their sizes and outlines and in the kinds of plate boundaries they possess, but actually there are a few simple observations that apply to all of them. The common point of meeting of three plates is called a *triple junction*. Although in abstract theory four plates could share a common point—like the Four Corners common point of Utah, Arizona, New Mexico and Colorado—the physics of rupture of a thin brittle plate makes a quadruple junction highly improbable.

Figure A shows six common kinds of plate junctions consisting of the three kinds of plate boundaries: converging, spreading, and transform. Note that the legend of the diagram corresponds with the map legend of text Figure 12.15. (The continental suture can be included within the converging boundary class.) What is added on the accompanying diagram is a system of broad, open arrows to show relative plate motions. ("S" means the plate is stationary relative to the other two.)

Code letters are applied to each of the three boundary symbols:

R for "rift' codes for a spreading boundary,

T for "trench" codes for a converging (subduction) boundary, and

F for "fault" codes for a transform boundary.

Thus, each kind of triple junction can be expressed as a three-letter code. The six kinds we will refer to have the codes RRR, TTT, TTF, TFF, RTF, and RFF.

(1) Your assignment is to find on the world map of plates (text Figure 12.15), a clear representative of each type and enter the information on Figure B. Write the letter code beneath the drawing of the junction. Write in the names of the three plates involved in each. Draw arrows to show relative plate motions. One of the three plates can be designated as stationary (S), if that seems reasonable.

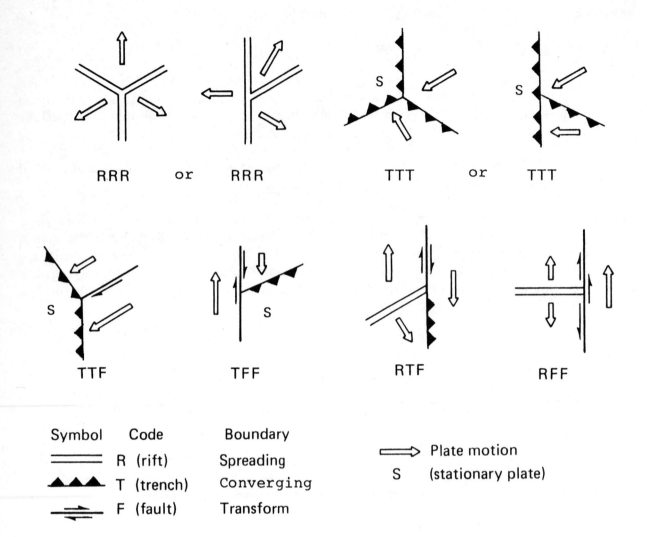

Symbol	Code	Boundary
═══	R (rift)	Spreading
▲▲▲	T (trench)	Converging
⇄	F (fault)	Transform

⇒ Plate motion

S (stationary plate)

Figure A

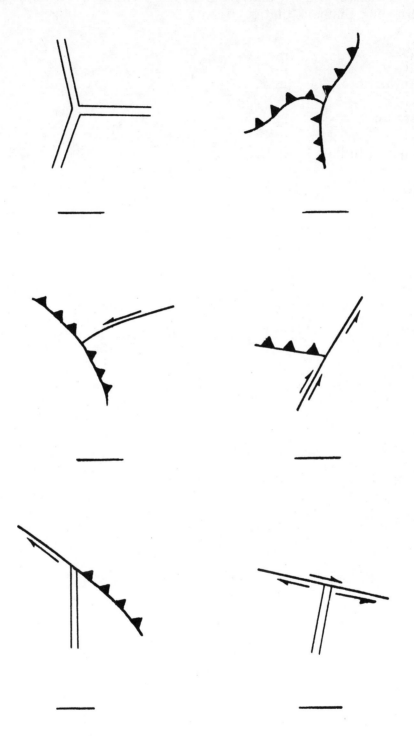

Figure B

(2) How many triple junctions does each of the following plates have? The meeting point of three colors on the world map, text Figure 12.15, will define a triple junction. (Disregard the Caroline and Bismark plates.)

Pacific   ——

Antarctic   ——

African   ——

Austral-Indian   ——

Nazca   ——

Cocos   ——

Caribbean   ——

Arabian   ——

Scotia   ——

Philippine   ——

Juan de Fuca   ——

Exercise 12-C Using Quasars to Tell How Fast the Plates Move

[Text p. 409–414, Figure 12.15.]

During the 1980s, the science of *geodesy* (from the Greek, "to measure the earth") took a giant stride ahead, thanks to special orbiting earth satellites that can be used to measure with great precision the surface distance between two points on the earth. A satellite known as LAGEOS was used to bounce back laser beams with such precision as to yield actual rates at which points on two different plates are either separating or converging. For a time, the measurement errors were so large, compared with the spreading or converging rates the data were yielding, that the results were in question. Then, in 1982, the measurement errors had been reduced to the point that plate movement rates were virtually a certainty within acceptable limits of error. Best of all, the observed motions were in the same directions predicted by plate tectonics and the rates were in the same ball park, at least.

Then came Very Long Baseline Interferometry (VLBI), making use of radio signals received from distant astronomical objects, such as quasars. To use these data, NASA had already set up an organization called the Crustal Dynamics Project, with participating observing stations in North America, South America, Europe, Australia, and Japan. By 1986, excellent VLBI results had been obtained, in close agreement with plate movement rates derived by geophysicists from the evidence of plate tectonics—paleomagnetic data, particularly. The data we will use are given in Table A. A plus sign means separation (spreading); a minus sign convergence (closing).

Table A VLBI Data of Relative Plate Motions

Stations	VLBI Rate cm/yr	Geophysical rate, cm/yr
Westford, CT, and Onsala (Göteborg), Sweden	+1.7	+1.7
Kauai, HI, and Fairbanks, AL	−3.9	−5.0
Kauai, HI, and Tokyo, Japan	−8.3	—
Perth, Australia, and California (N. American plate)	−7	—
Perth, Australia, and Nazca, Peru	+2 to +3	—
Vandenberg A.F.B., CA, and Fairbanks, AL	−7.9	−5.2

[Data sources: W.E. Carter and D.S. Robertson, 1986, *Scientific American*, vol. 255, no. 5, p. 46–54; Research News, *Science*, 1987, vol. 236, p. 1425–26.]

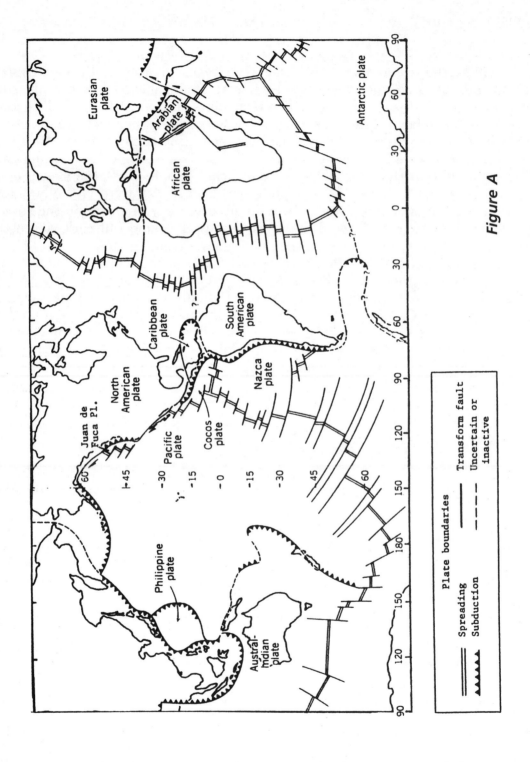

Figure A

Plate boundaries

Spreading   Transform fault

Subduction   Uncertain or
              inactive

(1) You are asked to plot the data of the Table A on the world map of lithospheric plates. We have provided a special copy of text Figure 12.15, on which you can insert data called for in the questions. On this map, Figure A, locate and label the following items of information:

Observing stations at the ends of each line. (Use first initial of the name.)

Connecting straight line between pairs stations given in Table A.

VLBI rate of relative motion (beside line).

Geophysical rate, in parentheses, near VLBI rate.

(2) Westford, CT, and Onsala, Sweden. (Use Göteborg as the equivalent location of Onsala.) What two plates are in relative motion on this line? Does this rate compare favorably with the long-term geologic rates generally given for the North Atlantic spreading rift? (To assist you in answering this question, we have inserted on the map, Figure A, three spreading rates along the Mid-Atlantic spreading boundary.)

_____

_____

_____

_____

_____

(3) Kauai, HI, and Fairbanks, AL. What plates are involved in this motion? What kind of plate boundary separates these plates? What relative motion and rate are observed on this line?

_____

_____

_____

_____

_____

(4) Judging from the geographical position of the northern Pacific plate boundary in the Aleutian region from Alaska to Kamchatka, does the Pacific plate motion follow the straight line connecting Kauai and Fairbanks? Draw a few arrows to show your interpretation of the motion of the Pacific plate at this northern boundary.

_____

_____

_____

_____

_____

(5) Kauai, HI, and Tokyo, Japan. Is it reasonable that the rate of motion on this line is much higher (−8.3 cm/yr) than on the line to Fairbanks (−3.9 cm/yr)? Can you explain this discrepancy?

_____

_____

_____

_____

(6) Perth, Australia, and California. The California station is located on the North American plate, inland from the San Andreas Fault. The convergence measured is thus between the Austral-Indian plate and the North American plate. Is the converging rate of −7 cm/yr adequately explained by the plates and boundaries lying between the two stations?

_____

_____

_____

_____

(7) Perth, Australia, and Nazca, Peru. Nazca is located at lat. 14°S, near the Pacific coast. (Aim for the sharp bend in the coastline.) Perth and Nazca are obviously moving apart, and this is opposite to the closing motion of Perth and California. Study the plate boundaries along the track and offer an explanation of the observed separating rate of between +2 and +3 cm/yr. (On our map, we have inserted a separation rate of +10 cm/yr for the East Pacific Rise at about latitude 30°S.)

_____

_____

_____

_____

_____

_____

NAME _____ *DATE* _____

Exercise 13-A Using Graphic Scales to Measure Distances

[Appendix of this exercise manual, Figure A.8.]

On the following page we reproduce a portion of a modern large-scale topographic map. The three graphic scales are duplicated below the map; you may cut them apart into three separate strips for measuring distances on the map. To measure distance, place a division mark of the scale on one point, such that the zero point falls short of the second map point and the remaining distance falls within the subdivision scale at the left. The scale below shows how this is done:

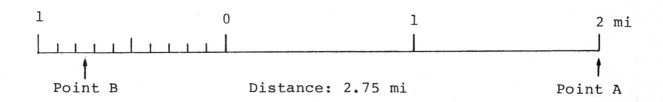

(1) What is the airline distance in kilometers between the Lighthouse and Santa Barbara Point?

_____ km

(2) What is the length in feet of the pier named "Stearns Wharf?"

_____ ft

(3) What is the distance in miles between the beacon on Point Castillo breakwater and Santa Barbara Point?

_____ mi

(4) Calculate the surface area in square kilometers within the limits of the map. (Measure length and width and multiply the two numbers.)

Length _____ km; Width _____ km; Area _____ sq km

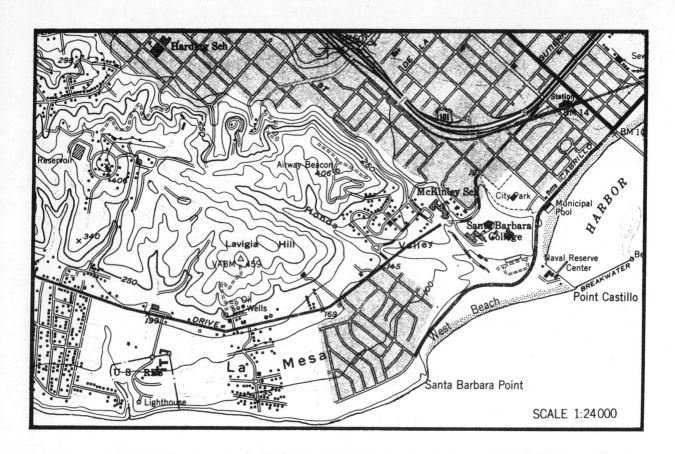

SCALE 1:24 000

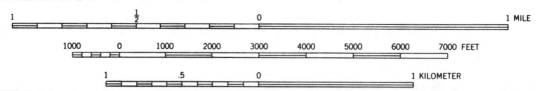

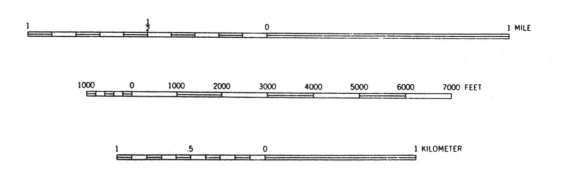

NAME _____   DATE _____

Exercise 13-B The United States Land Office Grid System

[Appendix of this exercise manual.]

Beginning in 1786, public lands lying west of the colonial states were surveyed and sub-divided into units called *congressional townships*, which are squares 6 miles on a side. Thomas Jefferson is said to have proposed this plan to Congress, and it was implemented under Thomas Hutchins, who had been appointed Geographer of the United States. At times, the surveyors' work had to be suspended because of danger of attack by hostile Native Americans. The Land Office grid system still dominates many cultural features of the landscape of the Middle West, the Great Plains, and valleys of the Far West. Patterns of highways, roads, city streets, farms, and counties follow this grid system..

Use as your text the *Appendix* of this exercise manual: See figures A.18, A.19, A.20, and A.21.

Figure A, below, is a portion of the Redfield, South Dakota, Quadrangle surveyed nearly a century ago by the U.S. Geological Survey. It shows sections of land and a standard parallel. Almost every section boundary is occupied by a road (double-line symbol) or a railroad. Convoluted lines are topographic contours and stream channels.

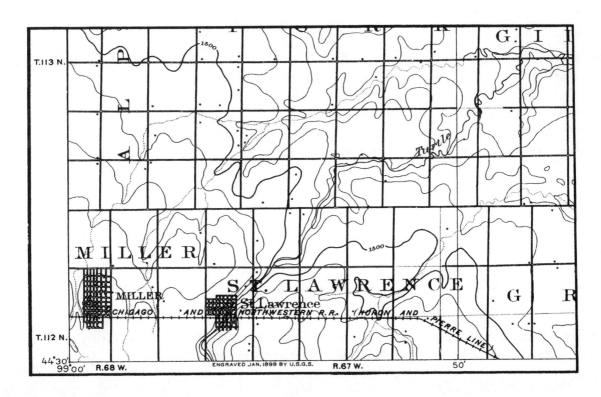

Figure A

(1) Find the township boundaries and highlight them with horizontal and vertical color lines. (Boundaries are shown by long dashes printed over the symbol for roads.) Why do you suppose the two towns—Miller and St. Lawrence—happened to grow up so close to each other?

_____

(2) On the map, number all sections with their correct numbers according to the established system shown in Figure A.19 in the Appendix.

_____

_____

_____

(3) Find and label the *standard parallel* on the map. Refer to Appendix Figure A.19. Can you explain why north-south section and township boundaries are offset along this parallel? Hint: Are meridians of longitude truly parallel to each other on the globe?

_____

_____

_____

_____

(4) Speculate as to why the sections in the tier located beneath the word "MILLER" (large bold letters) are more than one mile in north-south extent.

_____

_____

_____

(5) State accurately and fully the location of the letter "T" in the word "Turtle" on the map. (Refer to Appendix figures A.20 and A.21.)

_____

(6) Locate on the map each of the parcels of land described below. Carefully draw in the boundaries of each parcel and label with the corresponding letter.

    Parcel A: NE $\frac{1}{4}$ of Sec. 19, T.113 N, R.67 W
    Parcel B: N $\frac{1}{2}$ of SW $\frac{1}{4}$ of Sec. 35, T.113 N, R.67 W
    Parcel C: SW $\frac{1}{4}$ of NE $\frac{1}{4}$ of Sec. 13, T.112 N, R.68 W

NAME _____ DATE _____

Exercise 13-C Topographic Contour Maps

[Appendix of this exercise manual.]

The *Appendix* of this exercise manual, titled *Topographic Map Reading*, should be used as your text for this exercise, since the subject is not covered in your textbook, *Introducing Physical Geography*. Begin your preparation for this exercise by reading Appendix pages 359–361.

The topographic contour map is a special kind of isopleth map. As stated in Exercise 3-E, isopleths of elevation of the land surface are called *isohypses*. The U.S. Geological Survey publishes topographic contour maps on various scales covering the United States. These maps are bought in large numbers by hikers and campers, who learn to use them as an aid in following roads and trails in recreational areas and for taking off across country in wilderness areas lacking in any trails or guideposts.

In physical geography, the topographic contour map is used to display scientific information in quantitative form. That information depicts landforms, which can be studied and compared as a means of identifying and classifying them, and for developing and testing hypotheses about the processes that shape landforms.

In the next several chapters, we use topographic contour maps to represent many kinds of landforms, along with photographs and diagrams that help you to visualize a landform from the contours alone. First, however, you need to develop your skills in drawing contours and learning how they provide information on the configuration of the land surface.

EX. 13-C

A Contour Map of an Imaginary Island In the blank space provided below, draw a contour map of a semicircular island, the straight side of which has a very steep slope as compared with the remainder of the island. The island is about 5 mi wide and 6 to 7 mi wide. Use a map scale of one inch to one mile. The summit point of the island is 105 feet above sea level. The contour interval is 10 ft. Two stream valleys extend down the gently sloping sides of the island.

Contour Map of an Imaginary Island

Determining Elevations from a Contour Map Figure A is a contour map constructed to give you practice in estimating the surface elevation of points in various locations. Be sure that you are familiar with Appendix Figure A.7 and the text that accompanies it. Note that in two places contours that form a closed loop are distinguished by short, straight lines at right angles to the contour line; these are *hachured contours* and indicate *closed depressions*, which are hollows in the surface.

(1) What contour interval is used on this map? _____ ft

(2) Give the elevation of the contour at point A. _____ ft

(3) Give the elevation of the contour at point B. _____ ft

(4) Estimate the elevation of summit point C. _____ ft

(5) Estimate the elevation at point D. _____ ft

(6) Give the elevation of the hachured depression contour at E. _____ ft

(7) Estimate the elevation of the bottom of the depression at point F. _____ ft

(8) Estimate the depth of the depression at F below the lowest point of the rim that encloses it. _____ ft

Figure A

Drawing Topographic Contours On Figure B, complete the drawing of contours at intervals of 100 ft. Use a soft pencil and draw lightly at first. Then go over the lines in heavy pencil or ink.

(9) How high is the cliff at Point C in the southeastern part of the map?

_____ ft

(10) Determine the average slope of the ground, in feet per mile, along the profile line between the 200 ft and 500 ft contours in the northeastern part of the map. (Two bold dots show the profile segment to be measured.)

_____ ft per mi

Drawing a Topographic Profile Construct a topographic profile along the line A-B of the contour map, Figure B. Use the graph provided (Graph A). Be sure you first read pages 362–363 of the Appendix and understand the procedure illustrated in Appendix Figure A.9. Fold under the left side of the page along the dashed line, and temporarily attach the folded edge to the profile line A-B on the map. Use a vertical exaggeration of five times the horizontal scale. Label elevations at 100-ft intervals on the vertical scale.

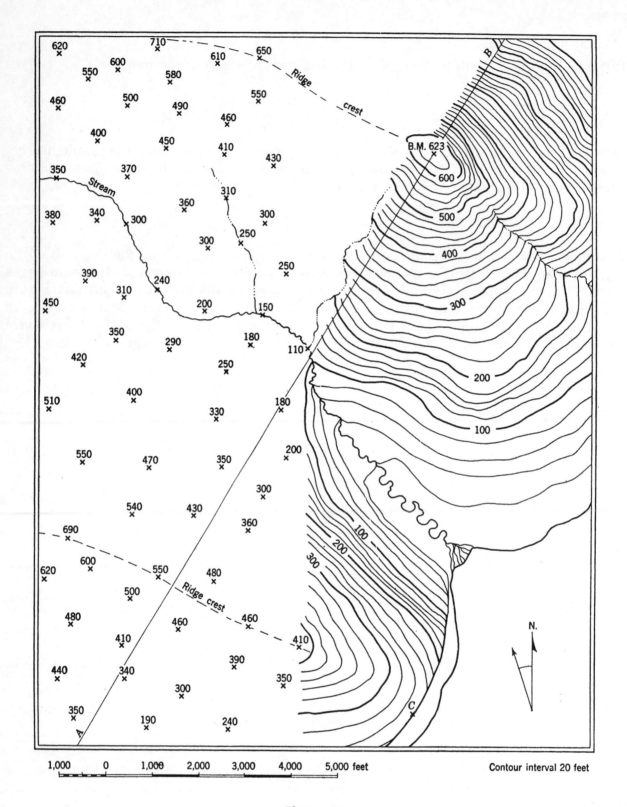

620
550 600
460 500 580 610 650
400 490 550
350 460 450 430
370 410
Stream 310
380 340 360 300
300 250
300 250
390 240 150
450 310 200 180
350 290 110
420 250 180
510 400 330 200
550 470 350 300
360
540 430
690 200 100
600 550 480 300
620 500 460 460 410
480 460 390
410 350
440 340 300 240
350 190
B.M. 623
600
500
400
300
200
100
100 200 300
410
C
N.

1,000   0   1,000   2,000   3,000   4,000   5,000 feet

Contour interval 20 feet

Figure B

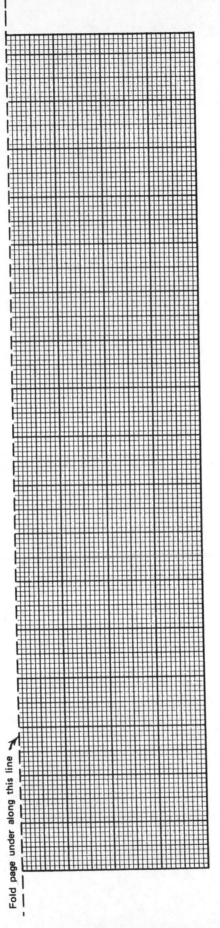

Graph A

Fold page under along this line

NAME _____   DATE _____

Exercise 13-D Mount Shasta—A Stratovolcano

[Text p. 427–430, Figures 13.2, 13.3.]

Volcanoes of the Cascade Range of northern California, Oregon, Washington, and south-ern British Columbia provide splendid examples of stratovolcanoes, two of which have de-livered major eruptions in this century. Greater and more devastating volcanic eruptions have occurred elsewhere in the same period, but the eruptions of Mount Lassen (1914–1915) and Mount St. Helens (1980) provide good examples of intense volcanic erup-tions and their environmental consequences. Included in the Cascade chain is a great caldera, Crater Lake, relict of a former lofty stratovolcano that has been named Mount Mazama. In this exercise, we study the great Cascade volcano, Mount Shasta, in north-ern California. It is a towering and strikingly beautiful landform.

Our Grid System In this and following exercises on landforms, our data sources con-sist not only of photographs but also of topographic contour maps, most of which are selected portions of topographic quadrangles published by the U.S. Geological Survey. To assist you to find and refer to designated features on these maps, we use a special grid system with 1000-yard units, scaled along the left and bottom margins of each map. This is a simplification of the UTM Grid System described in the Appendix of this manual (see pages 365–366). We will give the locations of features in *grid coordi-nates*, as illustrated in Figure A. The lower left-hand corner has the grid coordinates 0.0–0.0. The first number gives the distance to the right, the second number the dis-tance upward. Keep this in mind by using the slogan: "Read right up." For example, Point *A* has the grid coordinates 0.7–0.8, meaning "Right 700 yards, up 800 yards." Many of the maps also show a graphic scale of miles. Contour elevations are given in feet and it is up to you to determine the contour interval.

Nearly all topographic maps currently sold by the U.S. Geological Survey give contour in-tervals and elevations in feet. Why don't we use meters and kilometers as our units of grid measurement and elevation? Most Americans continue to use the English units of length and distance measurement with which they are thoroughly familiar. These allow us to re-late the dimensions of landforms to real-life experience. Consider that the radio messages of U.S. Navy fighter aircraft pilots, warning of the headon approach of hostile fighter planes, state the closing distance in miles! Need we say more?

Figure B is a photo of Mount Shasta, taken from a ground viewpoint on the southwest side of the peak, along a line from 0.10–0.0. Using infrared film, it shows distant features in great detail. The blue sky appears black, clouds and snow are white. The main part of the volcano is dissected (carved up) by streams and glaciers and no longer shows a crater. Several small glaciers, shown in gray overprint on the map, Figure C, remain on the moun-tain as relict features of the most recent (Wisconsinan) glacial epoch. A subsidiary vol-canic cone, named Shastina, is seen on the western slope of the main cone. Shastina is a relatively young feature and still shows a crater rim. Ski lifts and trails are located on the slope between Shastina and the main peak.

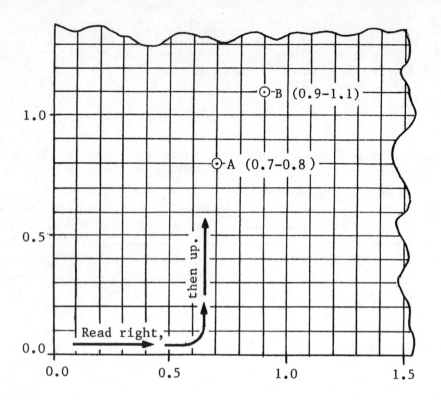

Figure A

(1) Give the grid coordinates of the summit of Mount Shasta. Estimate the summit elevation of this highest point.

Grid coordinates: _____

Summit elevation: _____ ft, plus _____ ft

(2) Estimate the summit elevation of Shastina, 13.5–17.7.

Elevation: _____ ft. plus _____ ft

(3) How wide is the volcano at its base, assuming the 5000-ft contour to represent the base? (Measure width on a line from 11-27 to 23-7.)

Width: _____ mi

(4) Calculate the angle of slope of the main volcano between the 10,000 and 12,000 foot contours. Take your measurement on the southwest side. On Graph A, plot the vertical and horizontal distances. Draw the hypotenuse of a right triangle to represent the average surface and measure its angle with a protractor.

Slope angle: _____ degrees

Figure B *Mount Shasta. (Infrared photograph by Eliot Blackwelder.)*

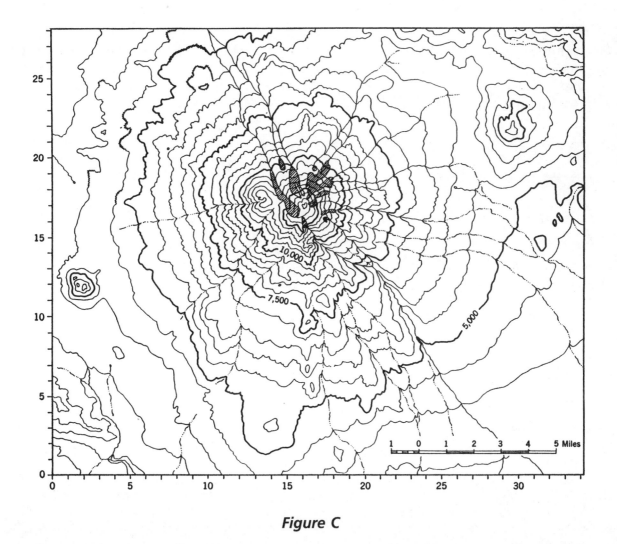

Figure C

(5) In the map area of 9.0–26.0 are special "sawtooth" contours, called *serrate contours*. They are symbolic of rough, blocky lava flows of recent date. Judging by the configuration of these contours, can you locate a likely point source for this area of fresh lava? (Give coordinates.) Find two additional areas of fresh lava. Label the three areas as "lava."

Source of lava flow: _____

Two other areas of lava flows: _____ and _____

(6) Describe the pattern of the streams that drain the slopes of Shasta. (Consult text p. 556 and Figure 17.22.)

_____

_____

_____

(7) Examine the round hill at 2.0–12.0. What is the origin of this feature? Look for it on the photograph. Are other similar features present in the eastern and southern areas of the map? Give coordinates. Look for these features on the photograph. Label these features on the map as "cone?"

_____

_____

_____

Special project:

(8) On blank Graph B, construct a topographic profile across Mount Shasta from 0.0–4.0 to 30.0–28.0. Calculate and label the vertical exaggeration of the profile.

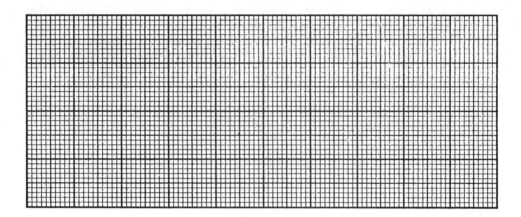

Graph A

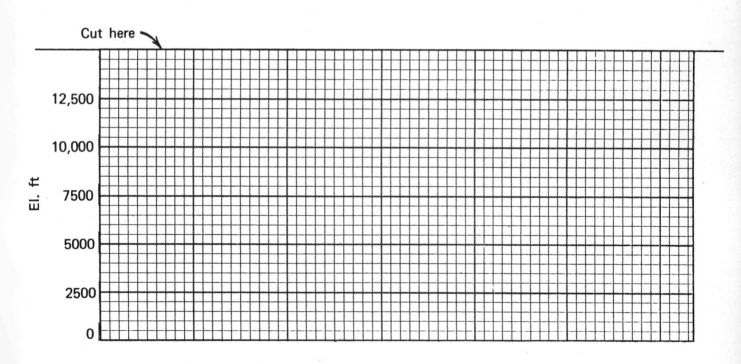

Graph B

NAME _____   DATE _____

Exercise 13-E Shield Volcanoes of Hawaii

[Text p. 430–433, Figure 13.5.]

The Big Island of Hawaii is endowed with the greatest active basaltic shield volcanoes on our globe. Iceland can also claim world-class standing in the competition for active basaltic volcanoes—among them Hekla, with 15 eruptions in historic time. Both Hawaii and Iceland are gigantic piles of basaltic lava built up from ocean floor depths of several thousand feet. But there is a difference: Whereas Iceland lies squarely on the Mid-Atlantic Ridge—the active spreading boundary between two plates—Hawaii and its older family members are far from any active plate boundary. As your textbook explains, they have been generated from a hot spot in the basaltic crust, thought to be fed by a mantle plume of rising magma that stays put while the Pacific plate rides over it, heading westward.

In this exercise, we focus on the summit region of Mauna Loa, largest of the active shield volcanoes. To investigate the rather strange volcanic landforms in this area, we make use of a splendid air photograph taken many years ago by what was then the U.S. Army Air Corps in the pre-World War II period (Figure A). A schematic block diagram (Figure B) and two geologic maps (Figures C and D) provide the information we need to interpret the photo.

(1) Using Figure C, name the active and inactive volcanoes shown on the map. What map evidence suggests that Mauna Kea is extinct, or at least, has long been inactive?

Active: _____

Inactive: _____

_____

_____

(2) Give the approximate summit elevations (ft) of Mauna Kea and Mauna Loa.

Mauna Kea: _____ ft. Mauna Loa: _____ ft.

Figure A

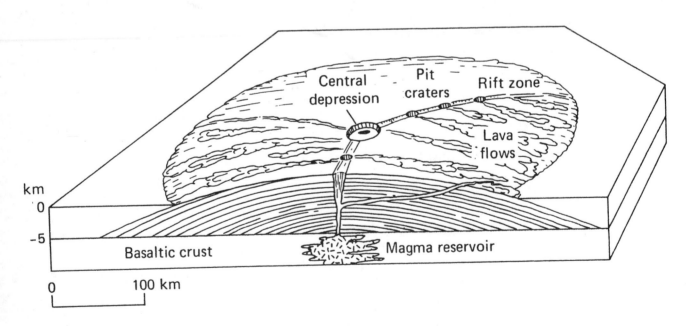

Figure B

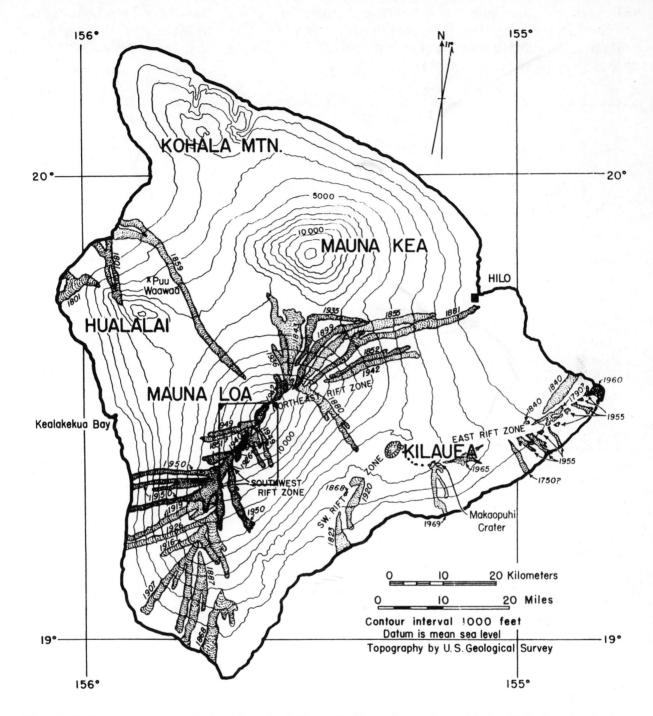

Figure C *Contour map of the island of Hawaii. (Based on data of the U.S. Geological Survey. From G.A. Macdonald and A.T. Agatin,* **Volcanoes in the Sea***. Copyright © 1970 by the University of Hawaii Press, Honolulu. Used by permission.)*

(3) Study the distribution of dates (years) given for the lava flows of Mauna Loa (Figure C). Note where the oldest flows are located and where the youngest occur. What general statement can you make about this distribution? What does it mean in terms of the recent history of volcanic activity?

_____

_____

_____

_____

_____

_____

Figure D shows some details of the summit region covered by the photograph. Note that the central depression is called a "caldera." Unlike the explosion caldera of Crater Lake, this depression results from collapse of the volcano summit as magma is withdrawn from below. The caldera floor may also be raised in level by the outpouring of magma.

(4) If the caldera and pit walls (steep cliffs) result from down dropping of solid lava, are these walls actually a kind of fault landform? If so, what kind of fault? (Consult text Figure 13.14.)

_____

_____

(5) On the southern boundary of the map, Figure C, mark and label a point that seems to lie beneath the aerial camera when the photo was taken.

(6) On the photograph, Figure A, write in the names of the three prominent pit craters and the caldera itself.

(7) What is the meaning of the long black lines on the map, Figure D? Identify and label one such feature clearly shown on the photograph.

_____

_____

_____

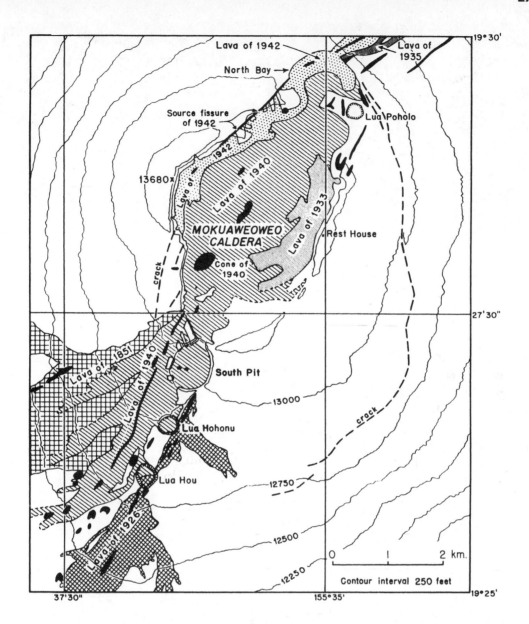

Figure D *Geologic map of the summit area of Mauna Loa, showing the lava flows of 1940 and part of the lava of 1942. (Same data source as Figure C. Used by permission.)*

(8) On the photograph, snow cover appears white. However, flows at the lower right, apparently coming from the area of Lua How, are black. Does this mean that the rock is so hot that it melts the snow?

_____

_____

_____

_____

NAME _____   DATE _____

Exercise 13-G Fault Scarps of the Klamath Lakes Region

[Text p. 437–441, Figures 13.14, 13.15, 13.16.]

The Basin and Range Province of Oregon, Nevada, and other states to the south and southeast is one of the major geomorphic divisions of the United States. You can think of it as a "pull-apart" section of the North American plate. Stretched out in an east-west direction, the brittle crust has fractured into many north-south normal faults, forming fault blocks. Some have remained high as horsts or tilted blocks; others have and foundered to form fault valleys and grabens. A striking example is the Klamath Lakes region of southwestern Oregon, just east of the southern end of the Cascade Range.

In this exercise we study the fault landforms of the Klamath Lakes region by using a contour topographic map (Figure A) in combination with a finely rendered block diagram by the distinguished geographer, Erwin Raisz (Figure B). Raisz was both a skilled cartographer and a free-hand landscape artist; his art shows in the softness and delicacy of his rendition of block diagrams, but allows little compromise with accuracy. We shall encounter more of his block diagrams in later exercises and in later chapter of your textbook.

(1) Using a red pen or crayon, draw on the map all fault lines you can confidently interpret from steep scarps shown by the crowding together of contour lines. Write the letters D and U on opposite sides of each fault to indicate downthrown and upthrown sides, respectively. Find and label an excellent example of a graben (G). Find and label an example of a horst (H). Give the locations of these two features.

_____

_____

_____

(2) What evidence can you give to conclude that the upthrown block at 20.7–7.0 has been tilted to the east?

_____

_____

_____

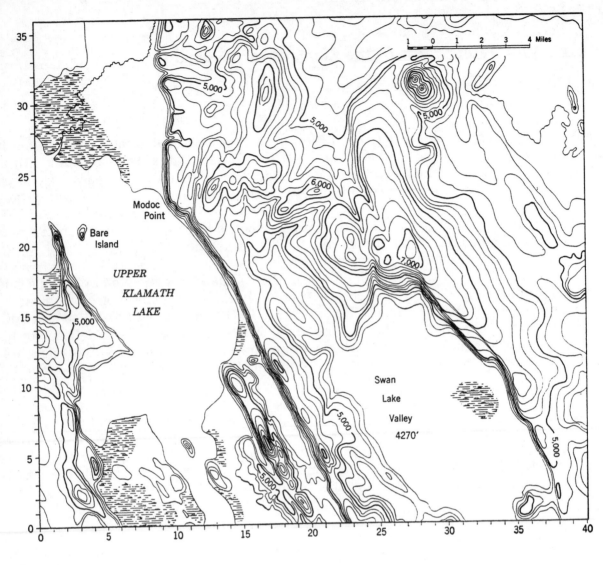

Figure A

(3) Is Swan Lake Valley a good example of a graben? Give evidence for your answer.

_____

_____

_____

_____

_____

(4) Estimate the height of the fault scarp at **(a)** 13.0–20.0 and **(b)** 35.0–10.0.

(a) _____ ft   **(b)** _____ ft

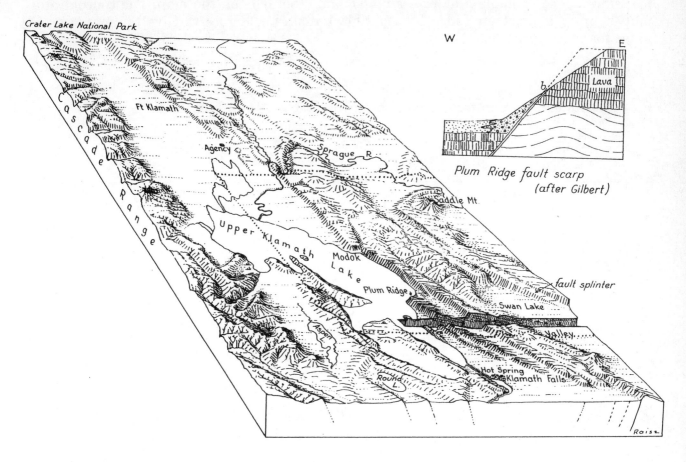

Figure B *Perspective diagram of the Klamath Lakes region, Oregon. A dotted line shows the limits of the accompanying topographic map. (Drawn by Erwin Raisz.)*

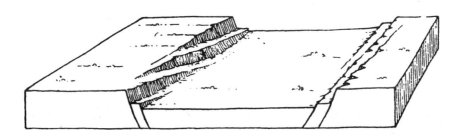

Figure C *A fault splinter consists of a narrow sloping fault block between two normal faults, one of which dies out as the other begins. Thus the total displacement is passed gradually from one fault to the other. Two splinters are shown here. (Drawn by Erwin Raisz.)*

(5) Find a good example of a *fault splinter* and label it FS on the map. The concept of a fault splinter is illustrated in the special block diagram, Figure C. Give the map coordinates for this feature.

Fault splinter: _____

(6) Offer an explanation for the flatness of the floor of Swan Lake Valley.

_____

_____

_____

(7) (Special problem.) Study carefully the contours just east of the fault scarp between 10-27 and 10-34. Describe what you find and relate the features you identify to the progress of up-faulting of the block.

_____

_____

_____

_____

Exercise 14-A Exfoliation Domes of the Yosemite Valley

[Text p. 460, 558–559, Figure 14.8.]

Deep mining for ores is hazardous work, made particularly dangerous by what miners call "popping rock." Without warning, from the ceiling and walls of a tunnel, or "drift," a massive slab of rock bursts free, falling on the unwary victims. A similar thing occurs in rock quarries, where the quarry floor may suddenly jump upward, as a thick slab of granite rifts loose. This phenomenon, which we call *unloading*, is caused by a spontaneous expansion of solid bedrock, previously somewhat contracted in volume under great pressure of overlying rock.

We include unloading under the heading of rock weathering, even though it is not caused by atmospheric agents, solar heat, or ice. In nature, unloading accompanies the overall process of landscape denudation, causing sheeting structure in massive rocks, such as granite, and often producing exfoliation domes. Nowhere are such domes more wonderfully displayed as landforms than in the Yosemite Valley of California, site of one of our oldest and most popular national parks. (See text Figure 14.8 and p. 558–559.)

We will investigate the Yosemite domes with the help of a photograph (Figure A) and a topographic contour map (Figure B). Two excellent specimens of exfoliation domes are shown: North Dome and Basket Dome. Label these two domes on the photograph.

Take some time to compare the map with the photograph, noting how the photo features are expressed in the contours.

(1) What contour interval is used on the map? _____ ft

(2) To get an idea how big these domes are, calculate the difference in elevation between the summit of Basket Dome and the floor of Tenaya Canyon.

Basket Dome summit _____ ft

Elevation of lowest contour in Tanaya Canyon _____ ft

Difference: _____ ft

(3) Lay out a trail from the floor of Tenaya Canyon to the summit of Basket Dome, using switchbacks (zig-gags) where needed, so that a rise of 100 ft uses no less than 300 ft of trail (horizontal measure). Start at point **P** on the stream. Draw your trail in pencil or pen directly on the map.

Figure A *North Dome and Basket Dome, viewed from the south. (Douglas Johnson)*

(4) Find the steepest slope shown on the map and mark it with a short line drawn across the contours. Determine the least horizontal distance of this line in a vertical distance of 400 ft (four contour intervals). Using the blank graph (Graph A), construct a triangle with these dimensions for legs and draw the hypotenuse. Using a protractor measure the angle that the hypotenuse makes with the horizontal. Enter your answer on the graph.

(5) Explain the succession of curious angular zigzag (sawtooth) bends in the contours at 0.5–0.4 and 0.2–0.2. Connect the nested zigzags in each group with a smooth curving line. Find these features on the photo and mark them with a color pencil or pen.

_____

_____

_____

_____

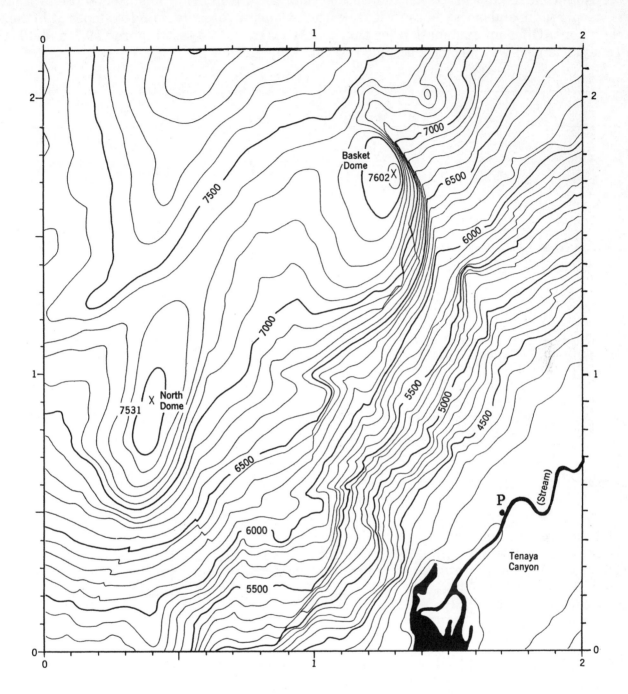

Figure B *Portion of the Yosemite National Park topographic map, scale 1:24,000, U.S. Geological Survey.*

(6) Find at least five other such groups of zigzag contour bends. Draw a color line along each. In the spaces below, give the map coordinates of each group.

————— ————— ————— ————— ————— —————

(7) Note that the topography of the wall of Tenaya Canyon below about 6500 ft elevation is steep, rough, and blocky in contrast to with the smooth, broadly rounded slopes at higher elevations. Offer an explanation for this contrast. (Hint: See text Figures 19.1 and 19.8. Tenaya Canyon was formerly occupied by an alpine glacier.)

_____

_____

_____

_____

_____

_____

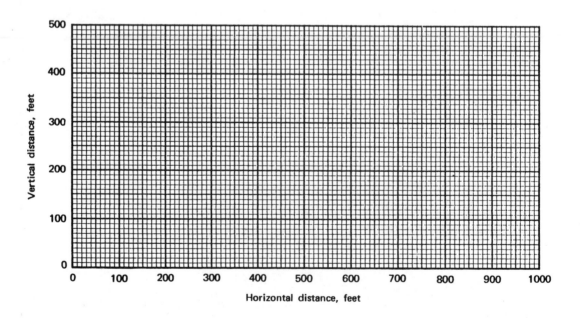

Graph A

NAME _____   DATE _____

Exercise 14-B The Great Turtle Mountain Landslide

[Text p. 468–469, Figure 14.22.]

The great Madison Slide, described in your textbook, was rivaled in size and destructive force by a Canadian landslide that occurred 53 years earlier at another place in the Rocky Mountains—in the southwestern corner of Alberta. Named the Turtle Mountain slide, after the mountain peak that broke free in 1906, it destroyed a large part of the town of Frank, which lay in the valley at the base of the mountain. Figure 14.22 of your textbook shows the Turtle Mountain landslide as it looks today. The accompanying drawing (Figure A), made from a photo taken shortly after the slide occurred, shows the mass of broken rock (left) and the great scar the slide left behind. The block diagram (Figure B) shows the scene from a different perspective. A detailed map of the landslide area (Figure C) was made during an investigation by geologists of the cause of the slide.

The great rock mass began its descent at a point over 3,000 ft (1000 m) above the valley floor. It quickly broke up into a rubble of huge boulders that roared across the channel of the Crow's Nest River. Engulfing part of the town, it killed and entombed about 70 persons. So great was the momentum of the mass that it traveled upgrade 400 ft (120 m) to reach a point about 1 mi (1.6 km) beyond the river bed. In terms of volume of rock in motion, that of the Turtle Mountain slide was about the same as for the Madison Slide, or perhaps a bit greater. Another section of the crest of Turtle Mountain (North Peak on the diagram) remains poised as a threat to the surviving part of the town. Similar geologic conditions favor another slide, which may come at any time.

Figure A *Sketch from a photograph taken shortly after the Turtle Mountain landslide. The view is toward the south from a high point to the right (north) of Frank, as shown in the block diagram.*

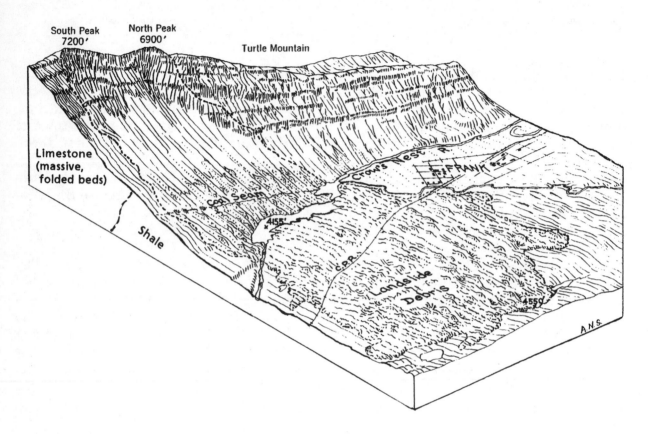

Figure B Block diagram of the area of the map, Figure C. (Drawn by A.N. Strahler.)

(1) Notice that different contour intervals are used on different parts of the map. Describe and explain this usage.

_____

_____

_____

(2) What is noteworthy about the appearance of the contours on the slide debris? Explain.

_____

_____

_____

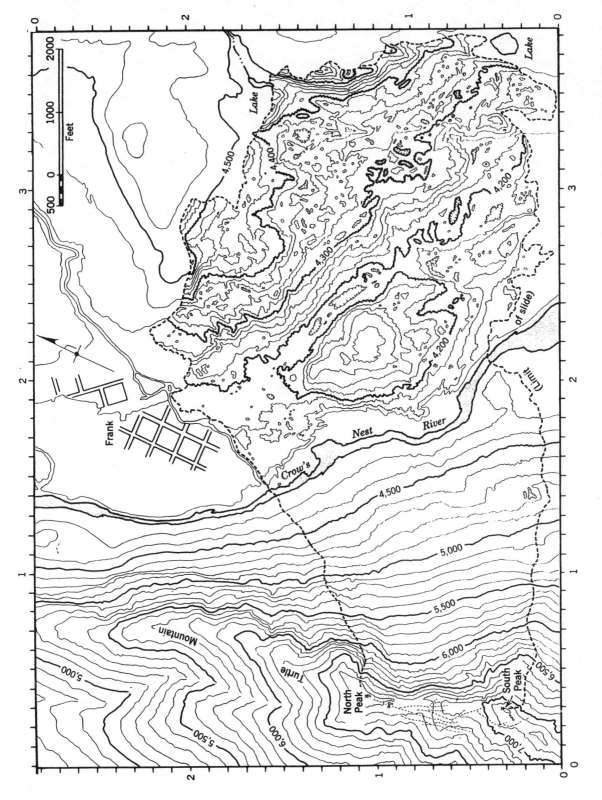

Figure C Contour topographic map of the Turtle Mountain landslide. (Canada Department of Mines, Geological Survey Branch, Memoir No. 27, 1911.)

(3) Numerous closed depressions, shown by hachured contours, are present in the area of slide debris. How did these features originate?

_____

_____

_____

_____

Special Project

(4) Attempt an approximate measurement of the average thickness of the landslide debris east of the Crow's Nest River. Take the volume of this debris to be 30 million cubic yards. Draw in the 500-yd grid lines and count the squares within the slide limit. Include fractions of part-squares around the edges. For a more accurate answer, use 100-yd grid squares. Show your data below.

Area of debris: _____ sq yds

Average thickness of debris: _____ yds; _____ m

Exercise 15-A The Annual Cycle of Rise and Fall of the Water Table

[Text p. 484–487, Figures 15.2, 15.3, 15.4.]

The Cape Cod peninsula juts into the Atlantic Ocean, its shape like a flexed Yankee arm with closed fist, boldly challenging any foreign invaders who might attempt a landing. Today, however, the Cape's threat is by Yankee invasion from the mainland, as each year developers build more houses to fill the demand for both year-around residence and summer use. The Cape's resident population in 1988 was up more than 500 percent from 60 years earlier, but the quantity of fresh ground water available remains just about the same. There is no other viable source of fresh water. What's worse, the existing supplies are prone to pollution from sewage and the leaching of toxic substances from town refuse dumps. Environmental planning to preserve this priceless resource requires scientific knowledge of the Cape's ground water system.

In this exercise we make use of the principles of ground-water recharge and the water table, covered in Chapter 15 of your textbook. We apply this information to the data of observation wells on Cape Cod. An observation well is one that is set aside to monitor the natural rise and fall of the ground-water table; it is located some distance away from other wells that are being pumped. The record of rise and fall of water level is what we interpret in this exercise.

Cape Cod is largely underlain by a thick layer of sand and gravel of glacial origin. The surplus water readily sinks into this porous substratum to become part of the water table. Flowing surface streams are generally absent.

Graph A shows the records of four observation wells maintained by the U.S. Geological Survey on Cape Code. Locations are shown on the accompanying map, Figure A. The record spans nine consecutive years, 1962–1970. The vertical scale is marked off in one-foot units. Included is a drought period that began in late 1963, intensified through 1965 and 1966, and ended in 1967.

(1) Using those parts of the graphs for the years 1962–63 and 1968–70 (before and after the drought), designate the typical month in which the annual peak occurs and the typical month of lowest level.

_____

_____

_____

_____

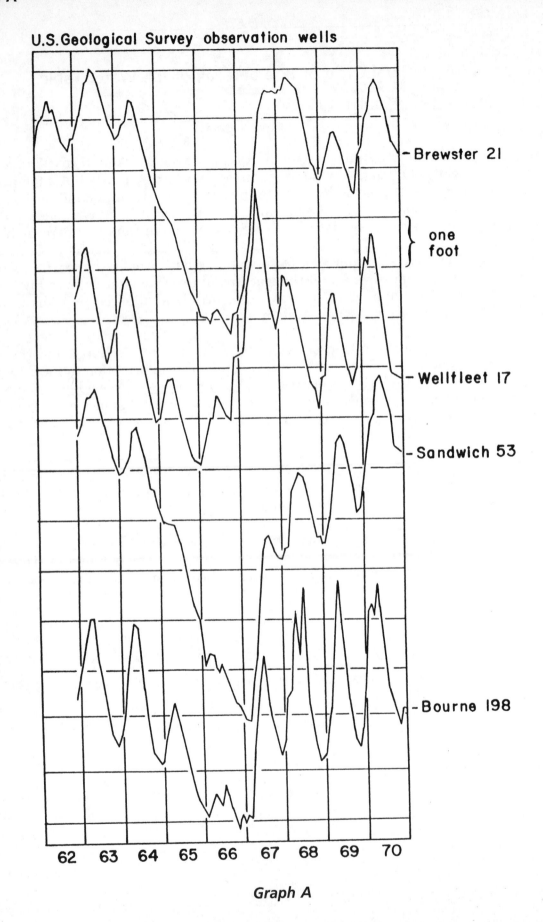

U.S. Geological Survey observation wells

Brewster 21

one
foot

Wellfleet 17

Sandwich 53

Bourne 198

62 63 64 65 66 67 68 69 70

Graph A

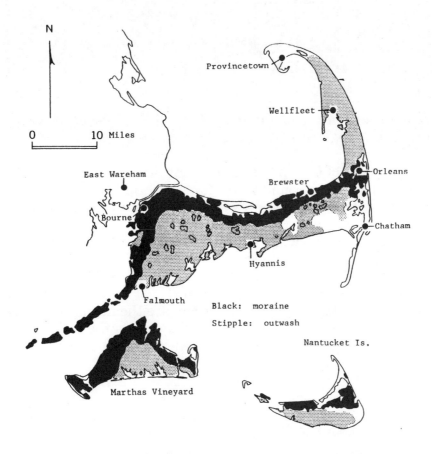

N

0   10 Miles

Provincetown

Wellfleet

East Wareham

Brewster

Orleans

Bourne

Chatham

Hyannis

Falmouth

Black: moraine

Stipple: outwash

Nantucket Is.

Marthas Vineyard

Figure A Map of the Cape Cod area, Massachusetts. (A.N. Strahler.)

Describing the annual cycle of fluctuation of the ground water table calls for an explanation in terms of the climatic factors that are involved. We turn next to three factors that will provide the explanation. Two are already familiar to you from Chapter 7. They are (a) the annual cycle of temperature and (b) the annual cycle of precipitation. The third factor (b) is the annual cycle of evaporation from the ground, including the foliage of all growing plants that may be present. Evaporation takes place from any moist soil surface, unless the soil water is solidly frozen. Evaporation from pores in the leaves of growing plants also takes place. This process of transpiration was explained in Chapter 9. There, you also learned that increase in air temperature causes an increase in the rate of transpiration.

Now, you need to evaluate these three factors throughout a yearly cycle for an observing station on or near Cape Code. We use the data of West Wareham, located only a few miles west of the Cape. Its climate is very similar in all respects to that of Cape stations such as Provincetown and Hyannis, but has a longer record. The table below gives mean monthly values of air temperature (T) in degrees Fahrenheit, precipitation (P) in inches, and combined evaporation and transpiration (E) in inches.

Data Table

	J	F	M	A	M	J	J	A	S	O	N	D	Year
T	29.0	29.1	37.3	47.3	57.7	66.6	72.6	71.2	64.3	54.4	43.9	32.7	50.5
P	4.30	3.59	4.80	4.28	3.45	3.26	2.88	4.29	3.84	3.44	4.60	4.20	46.85
E	0.00	0.00	0.35	1.26	2.91	4.17	4.80	4.41	3.27	1.89	0.79	0.00	23.85

Plot these three data sets on the blank graph (Graph B, lower part). Connect the plotted points with straight-line segments. Label each of the three plots.

On the upper blank graph of Graph B sketch a smooth curve of the ideal annual cycle of rise and fall of elevation of the ground water table. Be guided by the data of Graph A, selecting what seem to be the typical curves for normal years (1962–63 and 1968–70).

Compare your ideal curve with each of the plotted curves (T, P, and E) of the lower graph, looking for significant relationships to the ideal curve.

(2) Describe the annual precipitation regime. Compare the ideal ground water curve with your plotted line for monthly values of precipitation (P). Is the influence of precipitation a major or minor factor in determining how the water table changes in height?

_____

_____

_____

_____

_____

(3) Describe the cycles of both temperature (T) and evaporation (E) as to their form and timing. Is the relationship weak or strong? Explain.

_____

_____

_____

_____

_____

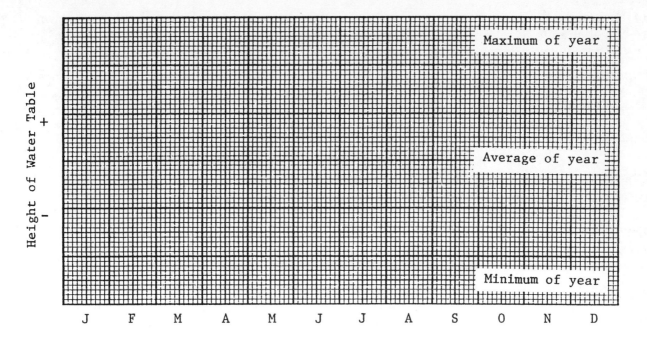

Idealized Cycle of Ground Water Fluctuation

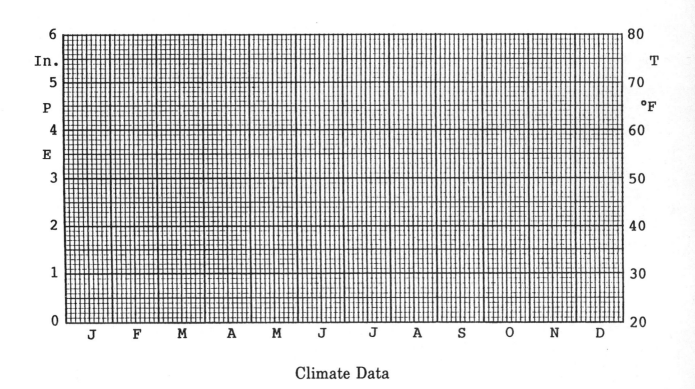

Climate Data

Graph B

(4) Compare the temperature and evaporation cycles with the ideal cycle of ground water level change. Explain the relationship you observe.

_____

_____

_____

_____

_____

_____

(5) During the drought period, what was the approximate total decline in water table level at each of the four stations? (Measure from 1963–64 minimum to minimum of lowest year of drought cycle.)

Brewster: _____ ft   Wellfleet: _____ ft

Sandwich: _____ ft   Bourne: _____ ft

(6) In the period 1968–70, the records of the four wells show some very marked differences in pattern. Bourne had three similar years; Sandwich rose in each successive year; Wellfleet and Brewster showed a dip in the maximum of the middle year. Can you suggest one or more possible causes for these differences?

_____

_____

_____

_____

_____

(7) Suppose that in the area surrounding an observation well, many new housing developments were added, each unit using a new well and disposing of its sewage in a septic tank with leaching field. What change might you expect to observe in the average level of the water table? Explain.

_____

_____

_____

_____

(8) Eventually, as new construction of homes continues, raw sewage returned to the ground water threatens to pollute that water source. To forestall this disastrous event, municipalities install sanitary sewer systems carrying the raw sewage to the coastline, where it is discharged into the ocean. What effect on the ground water table can you predict as a result of this outfall installation? Explain.

_____

_____

_____

_____

_____

(9) In what other way might the scenario you predict in Question 8 cause the ground water to become unpotable (not drinkable)?

_____

_____

_____

_____

NAME _____   DATE _____

Exercise 15-B Flood Regimes and Climate

[Text p. 499–502, Figure 15.23.]

In the summer of 1993, the mighty Mississippi River put forth one its great floods in its upper section of the Middle West, as if to make up for its performance in 1988, when it had suffered a great humiliation as it dwindled to record low stages. Commercial barge traffic came to a halt in some reaches of the river, seriously disrupting industrial activities dependent on cheap transport of bulk commodities. So, too little water in the channel of a large river can cause problems, though not comparable with the devastation of a major flood. In our modern industrial era, chemical pollution has taken away the romance of great rivers. Raw sewage and industrial wastes, dumped into the river, increase in concentration as the river stage lowers.

In this exercise we examine four strange-looking diagrams (Graph A), each telling the average discharge regime of an American river throughout the annual cycle. The measured quantity is stage (gauge height) in feet, which is directly related to discharge. Flood stage for each of these rivers at the location given is defined as the critical level above which inundation of the floodplain may be expected to set in (text Figure 15.13). However, like the exam grades of a college class, the frequency data are treated on a percentile basis. This form of presentation plays down the effect of differences in river magnitudes, allowing us to compare them in terms of response to the seasonal cycles of precipitation and runoff. Your knowledge of midlatitude climates will be put to use in analyzing these graphs.

The accompanying map of the United States, Figure A, shows the major rivers and their organization into 14 major regions or areas, according to the U.S. Geological Survey, as needed for reporting flow data in its Water-Supply Papers. With the aid of an atlas, locate the Lower Mississippi, Sacramento, Colorado (Texas), and Connecticut rivers. Use a color pencil or pen to highlight these rivers. Make a dot at the location of each of the four gauging stations and label with the name of the station.

Mississippi River at Vicksburg, Mississippi

(1) Describe in general terms the annual cycle of stages (gauge heights) at Vicksburg.

_____

_____

_____

_____

_____

_____

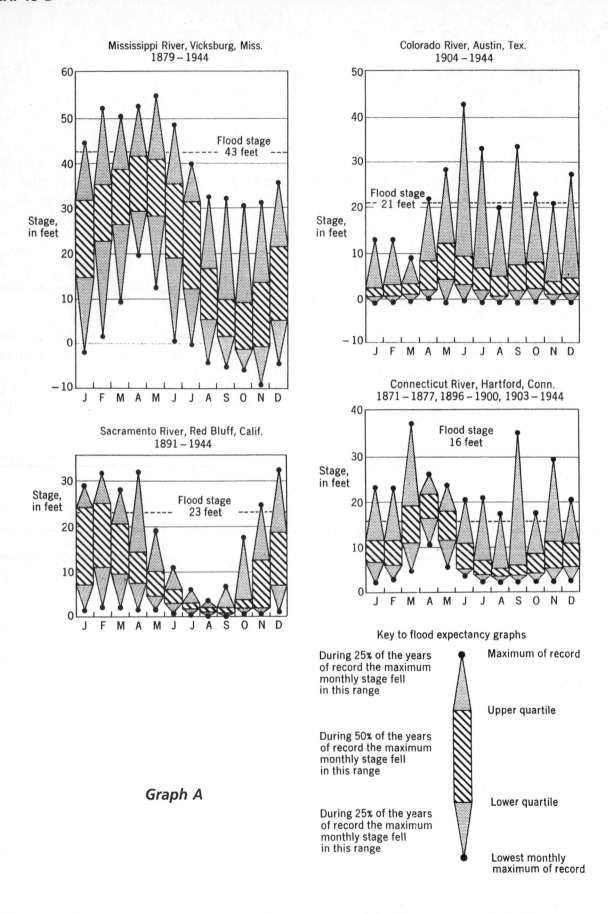

Mississippi River, Vicksburg, Miss.
1879 – 1944

Flood stage
43 feet

Stage,
in feet

Colorado River, Austin, Tex.
1904 – 1944

Flood stage
21 feet

Stage,
in feet

Sacramento River, Red Bluff, Calif.
1891 – 1944

Flood stage
23 feet

Stage,
in feet

Connecticut River, Hartford, Conn.
1871 – 1877, 1896 – 1900, 1903 – 1944

Flood stage
16 feet

Stage,
in feet

Graph A

Key to flood expectancy graphs

During 25% of the years
of record the maximum
monthly stage fell
in this range

During 50% of the years
of record the maximum
monthly stage fell
in this range

During 25% of the years
of record the maximum
monthly stage fell
in this range

Maximum of record

Upper quartile

Lower quartile

Lowest monthly
maximum of record

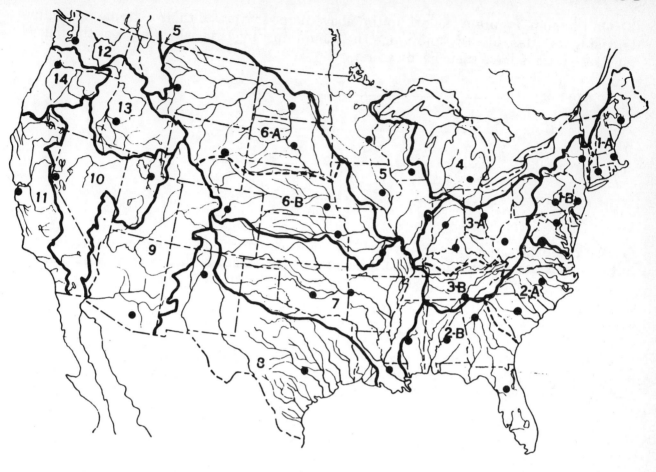

Figure A

(2) Study the entire Mississippi River system as shown in Figure A. The areas included are numbered 3, 5, 6, and 7. In the table below, enter the name of the principal river (or rivers) in each area or sub-area. Highlight these rivers in color on the map and label them. Identify and label those states, all or parts of which lie in these drainage areas.

Area	Principal River(s)
3-A	_____
3-B	_____
5	_____
6-A	_____
6-B	_____
7	_____

(3) On the map, Figure A, follow up the above-named rivers to their extreme upper limits. Using an atlas, identify and name the mountain ranges in which each of the following river systems have their headwaters:

Area 3: _____

Area 6-A: _____

Area 6-B: _____

Area 7: _____

(4) We can expect the release of runoff by snowmelt in the above mountain systems to have a major controlling effect on spring and summer stages of the lower Mississippi. Anticipate the order in which these runoff events would occur. Explain.

_____

_____

_____

_____

_____

_____

(5) Looking at the upper quartile and maximum of record, how do you explain the extreme lengthening of this quartile in September and October?

_____

_____

_____

_____

_____

Sacramento River at Red Bluff, California Using a good relief map of the western United States, examine the topographic features of northern California. (See also text figure 4.17.) Notice that Red Bluff lies near the northern end of the Great Valley and that tributaries to the Sacramento River extend north and east into the Sierra Nevada and the southern end of the Cascade Mountains.

(6) Describe the general features of the annual cycle for this river. How and why does it differ significantly from that for the Mississippi River?

_____

_____

_____

_____

(7) Examine the cycle of lowest monthly maximum of record. Describe and explain this cycle.

_____

_____

_____

_____

Colorado River at Austin, Texas Using an atlas, examine in detail the topographic features of the drainage basin of the Colorado River in Texas. Notice that the headwater drainage area is limited to the Llano Estacado, an elevated but flat part of the Great Plains. This high plain is separated from the Southern Rockies by the valley of the Pecos River.

(8) What are the most striking features of this graph? Explain.

_____

_____

_____

_____

_____

(9) How do you explain the extremely large maximum values occurring May through December?

_____

_____

_____

_____

_____

Connecticut River, Hartford, Connecticut Use an atlas to examine the long, narrow watershed of the Connecticut River, extending almost due north to the Canada border. Mountainous and hilly terrain, including the Taconic and Green Mountains, is typical of this region.

(10) Describe the annual cycle of the graph. Explain the salient features.

_____

_____

_____

_____

_____

(11) Suggest an explanation for the extremely high value of the maximum of record in September. Would a similar explanation hold for the exceptional maximum in March?

_____

_____

_____

_____

_____

_____

Exercise 15-C Evaporation from American Lakes

[Text p. 502.]

The need for great increases in supplies of fresh water is nowhere greater than in the arid climates, where irrigation agriculture is highly developed and urban centers are growing— that goes without saying. The catch is that loss of water by evaporation from lakes behind irrigation dams increases in proportion to their size and number. Engineers can easily design a dam so high, and with an area of water impoundment so vast, that it will never fill with enough water to supply downstream irrigation systems or to run the turbines installed in the dam to generate electric power. To say "never" may be an overstatement because, in due time, the blocked valley behind the dam will fill with sediment, reducing the water-storage volume sufficiently to allow the dam to fill, but that might take a half-century or more to come about. Small wonder the dam builders have been called "damn builders" by conservationists!

In this exercise, we study rates of evaporation from free water surfaces of lakes or ponds. Since evaporation is difficult to measure directly from such water bodies, hydrologists make use of the *evaporating pan*, which is simply a circular container 1.2 to 1.8 m (4 to 6 ft) in diameter and 25 cm (10 in.) or more in depth (Figure A). Water is added as required and the amount of surface lowering due to evaporation is measured with a gauge. Evaporation data are collected by the National Weather Service at many observing stations. Evaporation from the pan is generally greater from that of a lake or reservoir at the same location, so the pan readings require correction by a reduction factor ranging from 60 to 80 percent.

Figure A *An evaporating pan with anemometer at side. Notice a thermometer immersed in the water (upper left). The cylindrical device at the near edge of the pan is a hook gauge to measure the height of the water level. (U.S. National Weather Service.)*

Table A gives total annual evaporation, E, in inches for some actual reservoir surfaces. Also given are mean annual air temperature, T (°F), and mean annual relative humidity, RH (%). (Wind speed is another important factor, but we will not deal with it in this exercise.)

Table A

Code		E (in.)	T (°F)	RH (%)	Climate symbol	Köppen symbol
IT	Ithaca, NY ($42\frac{1}{2}°$N)	23	47	78	——	——
FA	Fallon, NV ($39\frac{1}{2}°$N)	57	51	52	——	——
BI	Birmingham, AL ($33\frac{1}{2}°$N)	43	63	72	——	——
EL	El Paso, TX (32°N)	71	64	36	——	——
SA	San Juan, PR ($18\frac{1}{2}°$N)	55	78	78	——	——
GA	Gatun, CZ (9°N)	48	80	84	——	——

(Data of National Weather Service.)

The stations are arranged in order of decreasing latitude. The first two make a northerly U.S. latitude pair; the second two a southerly U.S. pair. The last two, San Juan and Gatun, represent the tropical and equatorial latitude zones, respectively.

(1) Using the world climate map, text Figure 7.2, find these six stations and determine the climate of each. Enter the climate symbol in Table A. Space is also given for the Köppen climate symbol (see text p. 222–225).

(2) The accompanying U.S. map, Figure A, shows isopleths of annual evaporation from shallow lakes. Locate the U.S. stations on the map, marking each with a dot and labeling it with the station code.

(3) What meteorological factors are responsible for the low value of E for Ithaca?

_____

_____

_____

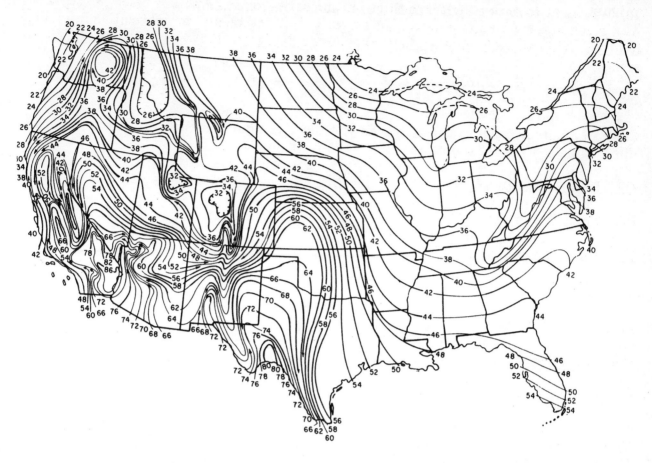

Figure B *Average annual evaporation (inches) from shallow lakes. (U.S. National Weather Service.)*

(4) Fallon has about 2.5 times the evaporation of Ithaca. Explain why this should be so.

_____

_____

_____

_____

_____

(5) Why is Birmingham's evaporation about double that of Ithaca?

_____

_____

_____

(6) Why is evaporation higher at El Paso than at Birmingham?

_____

_____

(7) San Juan, although warmer than any of the U.S. mainland stations, has an intermediate value of E, about the same as for Fallon. Explain.

_____

_____

_____

_____

(8) Gatun's evaporation is even less than that of San Juan, and not much different from
that of Birmingham. Explain.

_____

_____

_____

_____

(9) Where in the U.S. is evaporation least? Give both location and climate type.

_____

_____

_____

(10) What is the highest value of E shown on the U.S. map? With what geographical region is this high associated?

_____

_____

_____

NAME _____   DATE _____

Exercise 15-D Sinkholes of the Kentucky Karst Region

[Text, p. 488–489, Figures 15.7, 15.9, 15.10.]

The karst landscapes of Kentucky and southern Indiana have been developed in a moist continental climate in areas where thick limestones of Mississippian geologic age lie at or close to the surface. Mammoth Cave, established as a national park in 1936, lies in south-central Kentucky, a few miles northeast of Bowling Green. It is one of the largest cavern systems in the world, with about 150 miles (240 km) of known passageways. Five separate levels of passageways have been formed, the lowest of which carries an underground river, Echo River, that drains into the Green River. Kentucky pioneers discovered the cave in 1799, and during the War of 1812 saltpeter deposits formed from bat dung were mined to make gunpowder.

Caverns don't show on topographic maps, but their presence at depth is often indicated by a karst landscape with numerous sinkholes. In this exercise we examine a small section of a contour map in an area not far from Mammoth Cave National Park. We use another of Erwin Raisz's fine block diagrams for guidance in interpreting the map. It illustrates the sinkhole plain located south of Mammoth Cave and long known as the "Pennyroyal." (Pennyroyal is an aromatic American mint with blue or violet flowers.) The small portion shown in Figures A and B is typical of this plain.

(1) What contour interval is used on the map, Figure B? (Refer to the numbered heavy contour and to the spot height at 6.5-4.4.)

      Contour interval: _____ ft

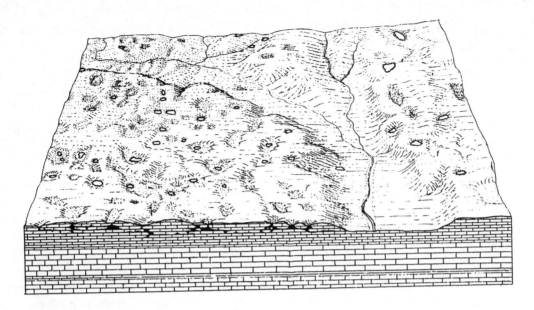

Figure A *Block diagram of the area shown in Figure B (Drawn by Erwin Raisz.)*

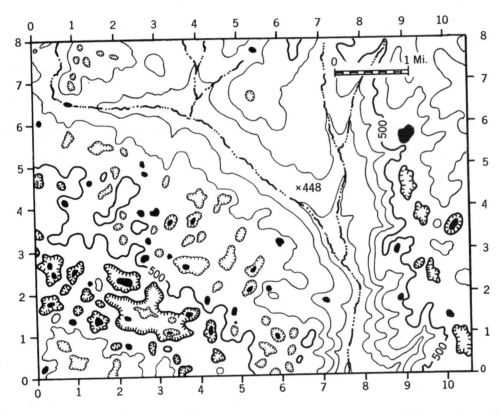

Figure B *Contour topographic map of a portion of the Sinkhole Plain of southcentral Kentucky. (Princeton, Ky., Quadrangle, 1:62,500, U.S. Geological Survey.)*

Depth of a sinkhole can be estimated from its depression contours. By "depth" we mean the difference between the lowest outlet point on the rim of the depression and the deepest point on the bottom of the depression. To illustrate this concept, the profile diagram below may be of help.

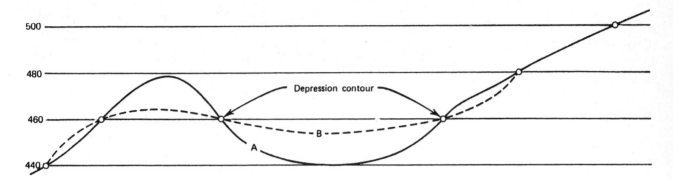

Given one set of contours, the true profile may approach Profile A (maximum depth) or Profile B (minimum depth). Thus we can only say for sure that the depression depth (d) is more than zero but less than 40 feet (two contour intervals). So we write: $0 < d < 4$

(2) Estimate the depth of each of the sinkholes for which the location is given below by grid coordinates.

Grid coord.	Depth range
2.7-1.5	_____
2.2-2.3	_____
9.1-4.5	_____

(3) On the broad divide located at 6.5-5.5, draw contours to show a sinkhole with depth more than 40 ft but less than 80 ft. Label the elevation of the outermost contour.

(4) (Extra credit) Using the blank graph provided below, construct an east-west profile from 0.0-1.6 to 10.6-1.6.

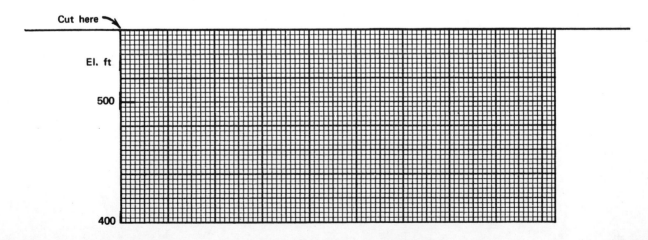

NAME _____   DATE _____

Exercise 16-A Niagara Falls

[Text, p. 520–526, Figures 16.9, 16.10, 16.11, 16.16, 16.17.]

Waterfalls are young landforms with short life spans. If Nature abhors a vacuum, Nature also abhors a waterfall. The enormous concentration of kinetic energy at the brink of a fall guarantees that rapid lowering will occur, unless exceptional geological conditions are present. In this exercise, we examine Niagara Falls. Its two parts—the Canadian, or Horseshoe, Falls and the American Falls—dominate the viewer by sheer power and enormous water volume. The falls were formed in late Pleistocene time, and emerged as a product of both glacial ice and crustal tilting. Our emphasis here is on these falls and their spectacular gorge as landforms produced by stream erosion.

The two falls of the Niagara River are seen in an oblique air photograph (Figure A). Compare the features of the photograph with those shown in the text diagram, Figure 15.15. The topographic contour map, Figure B, shows the entire stretch of the Niagara River from above the falls to where it emerges from the Niagara Escarpment. Compare the map area with the block diagram, text Figure 16.16.

Figure A *Oblique air photograph of Niagara Falls. (Geological Survey of Canada, Atlas No. 101.)*

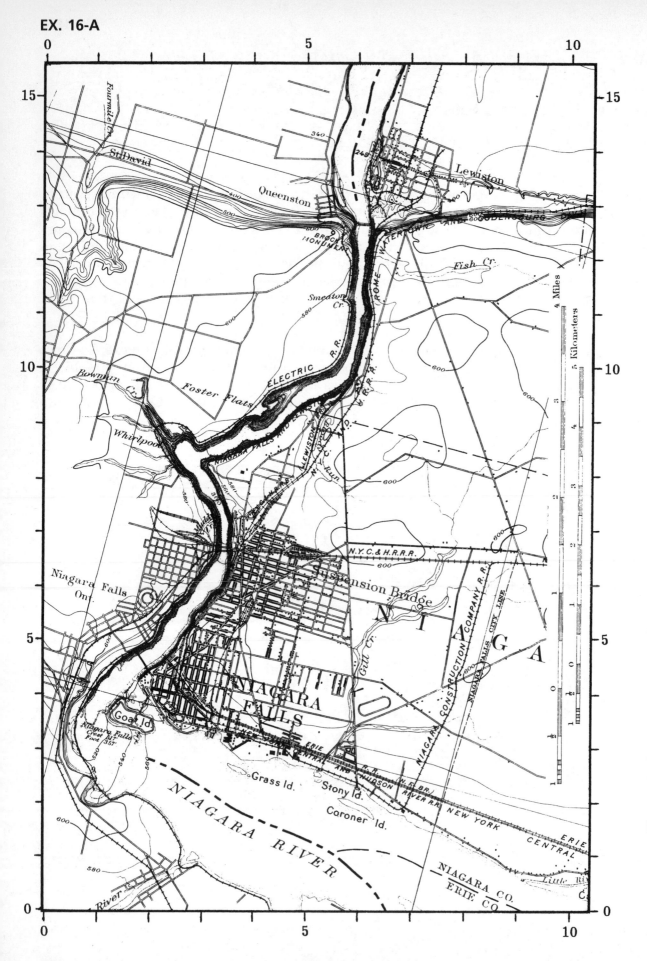

Figure B Contour topographic map of the Niagara Falls and the gorge of the Niagara River. Scale 1:62,500. (U.S. Geological Survey, 1986.)

Figure C is a set of two schematic cross-sectional drawings. The upper part (a) shows an idealized situation in which a high vertical waterfall is formed quickly by normal faulting across the path of a river. The rock within the cross section is assumed to be perfectly uniform in composition and hardness. As the fall retreats upstream, it is quickly transformed into rapids, and these are rapidly cut down and replaced by a smoothly descending graded profile.

The lower part (b) of Figure C shows the geologic features related to the Niagara River gorge and its falls. The major feature is a massive layer of dolomitic limestone of a Silurian formation called the Lockport Dolomite. Beneath this resistant cap is a weak shale formation—the Rochester Shale—and below that a succession of interbedded sandstones and shales. On either side of the Niagara gorge the Lockport dolomite forms a cuesta that drops off abruptly along the Niagara Escarpment. The dip of the strata is gently southward from the escarpment, so that as the falls retreat, their elevation becomes lower.

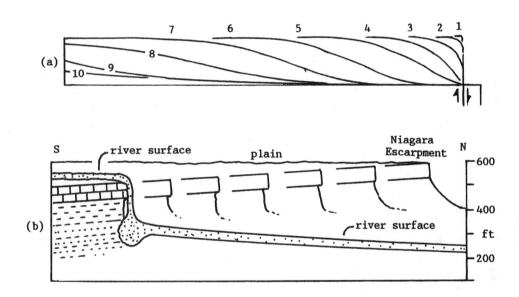

Figure C *Schematic cross section of the Niagara Gorge and waterfalls.*

The Horseshoe Falls Figure D reproduces an original drawing by the noted geologist, G.K. Gilbert, it shows his concept of the plunge pool of the Horseshoe Falls and the water currents responsible for eroding the base of the falls. (Note: Text Figure 16.17, drawn by Erwin Raisz, is in error for showing a deep plunge pool at the foot of the American Falls. Perhaps he did this because he couldn't show the plunge pool as lying at the foot of the Horseshoe Falls, where it belongs. Shouldn't we allow Raisz this bit of artistic license?)

The Horseshoe (Canadian) Falls, as its name implies, has a deeply concave outline and is located at the head of the gorge. Moreover, a deep plunge pool lies below this curved rock wall. The lip of the American Falls is more nearly straight across (disregarding a great chunk of the caprock that fell out in 1954) and lies along the side wall of the gorge. Besides that, instead of a plunge pool, a great accumulation of limestone blocks lies at the base of the American Falls.

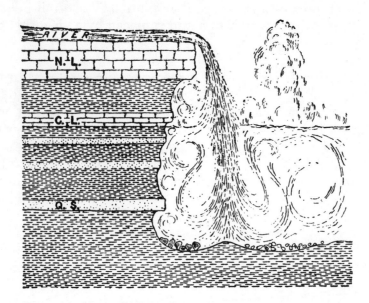

Figure D *A cross-sectional drawing of the Horseshoe Falls. (From G.K. Gilbert, 1896, Niagara Falls and Their History, in* Physiography of the United States, *American Book Co.)*

(1) Offer an explanation for these differences between the two waterfalls.

_____

_____

_____

_____

_____

_____

Rate of Retreat of the Horseshoe Falls In 1827, a British naval captain made a detailed drawing of the cataract at the lip of the Horseshoe Falls. He used a *camera lucida*, which provides a fixed frame to get an extremely accurate rendition of the details. His drawing is reproduced here as Figure E. In 1842, the New York State Geologist, James Hall, made instrumental surveys of the same scene, showing that substantial retreat had occurred. Other precise surveys followed. A recession of 220 ft had occurred between 1842 and 1896, for an average of about 4.6 feet per year.

Figure E *Captain Basil Hall's 1827 camera lucida drawing of the Horseshoe Falls, seen from Forsyth's Hotel. (Same source as Figure D.)*

(2) Assuming that the above calculated rate has held since the falls began at the Niagara Escarpment, calculate how long it has taken to form the entire river gorge.

Length of gorge: _____ mi, or _____ ft.

Total time for gorge to form: _____ years

(3) Radiocarbon dates of beach deposits at Lewiston establish the date of start of the ancestral Niagara Falls at the escarpment as −12,000 years. Using that figure, calculate the average annual rate of retreat for the entire gorge. Comment on the large discrepancy between this rate and the one based on recession of the lip of the falls.

Average rate of retreat: _____ ft/yr

_____

_____

_____

_____

_____

_____

_____

NAME _____   DATE _____

Exercise 16-B Alluvial Terraces

[Text p. 526–530, Figures 16.19, 16.20.]

Alluvial terraces made life a great deal easier for white settlers in New England and the Middle West, who established new towns and carved out new networks of roads and railroads connecting them. Advances and retreats of the great North American ice sheets in the Pleistocene Epoch had made things both worse and better for the newcomers. The bad news was a cover of bouldery glacial till that defied the plow on the uplands. The good news was the filling of valleys with alluvium, carved by post-glacial streams into strips of flat land running for tens of miles at a stretch along the valley sides of the major streams. These natural pathways came already graded and ready for roads and railroads. They also provided prime agricultural land and well drained sites for towns. As a bonus, there were many points at which rapids and falls in the stream bed could be put to use to turn the waterwheels that powered grinding and textile mills.

Alluvial terraces pictured in text Figures 16.19 and 16.20, belong to one major class, or variety, of alluvial terraces. It is a sequence of *unpaired terraces* (or "unmatched terraces") shown Figure A, upper part. The sequence of development of unpaired terraces is laid out in the lower diagram of Figure A. We call the flat land of the terrace itself the *tread*. The steep, undercut slope that bounds the tread is called the *scarp*. As in a staircase, each terrace tread is bounded on one side by a rising scarp and on the other by a descending scarp. Study the text photo of New Zealand terraces, text Figure 16.20. Identify as many treads and scarps as you can.

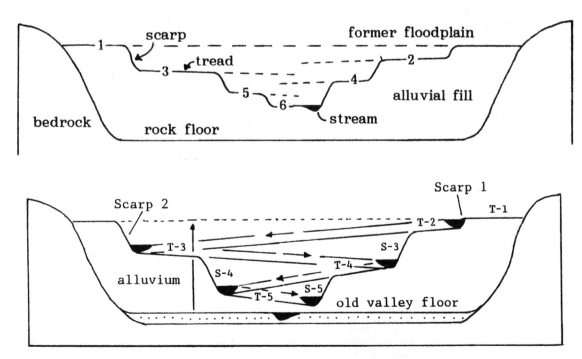

Figure A *(Above) Unpaired alluvial terraces. (Below) Stages in their formation.*

EX. 16-B

(1) The lower diagram in Figure A gives a detailed treatment of the development of unpaired terraces. Review and explain the plan of development shown here, starting with the former open valley with its bedrock floor.

Initial condition: _____

_____

_____

Aggradational event: _____

_____

_____

First downcutting event, including scarp and terrace formation: _____

_____

_____

_____

_____

(2) Text Figure 16.19, diagrams B and C, are reproduced here as Figure B. Identify and label the terraces and scarps in the manner shown in Figure A, (i.e., "T-1, T-2," etc. "S-1, S-2", etc.). Where space is tight, place the label outside the diagram and run an arrow to the feature.

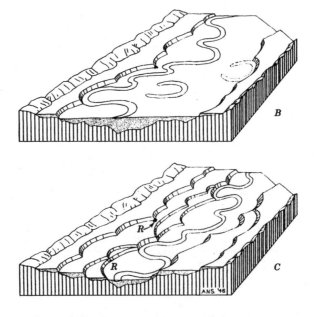

Figure B *Formation of unpaired alluvial terraces (A.N. Strahler.)*

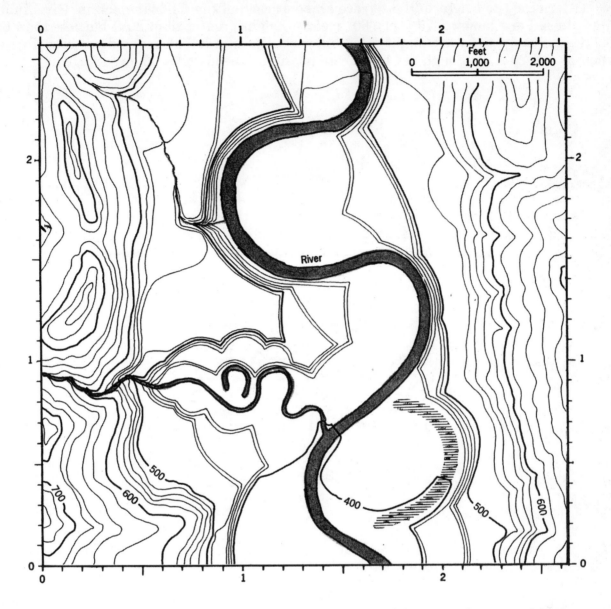

Figure C *Contour topographic map of imaginary alluvial terraces. (A.N. Strahler.)*

Figure C is a synthetic topographic contour map, not representing any real area, but serving as a diagram on which terrace features are idealized. It illustrates certain typical features of New England river terraces. The contour interval is 20 ft; the scale is 1:17,000.

(3) Determine the height of the terrace scarp at each of the following points. Give limiting values. For example, "20 < h < 60" means "height greater than 20 ft but less than 60 ft." (Conceivably, the scarp might extend upward almost a full 20 ft and downward almost another full 20-ft. Although not likely, the possibility needs to be allowed for.)

Scarp at 1.1-2.2: _____ < h < _____

Scarp at 1.6-1.7: _____ < h < _____

Scarp at 1.4-1.0: _____ < h < _____

Where terrace scarps are being carved by a stream that has well developed meanders, the scarps tend to take the form of *arcs*, concave toward the valley axis. These arcs are fitted to the size of the meander bend (more or less). Adjoining arcs are separated by sharp *cusps* pointed toward the valley axis. The accompanying map-diagram illustrates these terms:

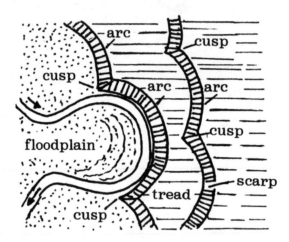

(4) On the map, Figure C, find and label three good examples of an arc, its two cusps, and a closely related meander bend.

(5) Study the terraces and scarps on either side of the tributary stream in the vicinity of 1.0-0.8. Note the close match between the size of the stream meanders and the size of the arcuate terrace scarps north and south of the stream. Then note the size of the meander bends of the large stream in comparison with the arcuate terrace scarps it has produced. Make a general statement to describe this relationship.

_____

_____

_____

_____

(6) What is the origin of the semicircular swamp extending from 1.7-0.2 to 1.8-0.8? Name this feature and label it on the map. Find and label another similar floodplain feature.

_____

_____

_____

(7) Why does the small stream have such a steep gradient at 0.5-0.9, but such a low gradient at 0.9-0.8.

_____

_____

_____

(8) Why does the contour line at 1.95-1.9 bulge westward in this place instead of bending east in a sharp **V** as the higher contours do just east of it? What landform is present?

_____

_____

_____

(9) If deep pits were to be dug in the vicinity of 1.0-1.2, what composition and texture of material would be revealed, and why?

_____

_____

_____

NAME _____   DATE _____

Exercise 16-C Floodplains and Their Meanders

[Text p. 521–522, 531–533, Figures 16.10, 16.21, 16.22, 16.23.]

Alluvial meanders, those snakelike bends of a river channel that require it to travel so long a distance to make its way to the sea, build the character of the entire floodplain. As meander loops grow, are cut off, abandoned, and replaced by new ones, the river shifts back and forth over the floodplain, constantly reworking the alluvium and shaping it into new landforms. As much as engineers have tried to eliminate those meander bends and replace them with straight reaches, they just won't stay straight. Nature establishes a steady river regime that, for all its continual change and seeming inefficiency, holds the total length of its meandering course more or less constant. Wouldn't it be better for humans to learn to live with that natural regimen?

A Floodplain in the Making We examine first a floodplain in a rather early stage of its development, much like that shown in Block C of text Figure 16.10, and in the final stage of Figure 16.13. Our contour topographic map, Figure A, is a portion of the Kanawha River valley in West Virginia. The floodplain is easily identified as a belt of land free of contour lines. It contrasts strongly with the surrounding hills, which rise to heights as much as 500 feet above the floodplain.

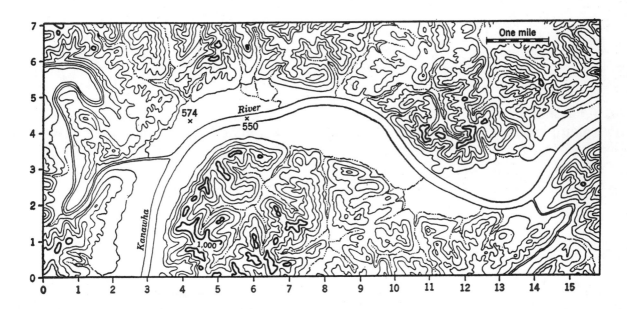

Figure A *Map of a portion of the St. Albans, WV, Quadrangle. Scale 1:62,500. (U.S. Geological Survey.)*

(1) Using a color pencil or pen, draw lines to show the limits of the floodplain. Label the floodplain. Measure the width of the river channel and of the floodplain. Make your measurements near the elevation number 574.

    River width: _____ yds

    Floodplain width: _____ yds

_____

_____

_____

_____

(3) From map data alone, can the gradient of the river be measured? Explain what problem is involved here.

_____

_____

_____

_____

_____

(4) (Optional) Compare the size of bends of the river channel with the size of bends of the floodplain, i.e., of the valley itself. What kind of stream activity is suggested by this relationship?

_____

_____

_____

_____

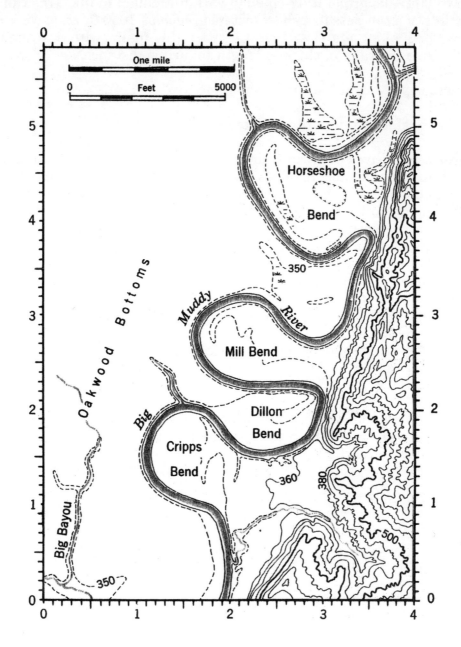

Figure B *Map of a portion of the Gorham, IL, Quadrangle. Scale 1:24,000. (U.S. Geological Survey.)*

Meanders on a Floodplain Fully developed meanders on a broad floodplain are illustrated in the accompanying map, Figure B, of the Big Muddy River in Illinois. The channel lies close to the valley bluffs on the east, while to the west there extends a broad, flat floodplain, Oakwood Bottoms. The river channel is shown by the shaded band. An additional contour (350 ft) brings out details that would otherwise be lost. (Note also that three 20-ft contours have been omitted between 380 and 500 ft.) The gradient of Big Muddy River cannot be calculated from this map alone, but is on the order of 2 ft per mile, or less.

(5) Measure the total length of river channel within the map limits. This can be done by setting the points of a compass to a short distance, such as 1000 ft, and swinging the compass back and forth along the river line. Compare this figure with a straight line from start to finish. Give the ratio of channel length to airline distance.

Channel length: _____ ft. Airline distance: _____ ft.

Ratio, channel length to airline distance: _____ to 1

Note: The Mississippi River between Cairo, IL, and Baton Rouge, LA, held to a more or less constant river length of about 850 mi from 1765 to 1930, despite having 19 major cutoffs and many minor channel shifts. For an airline distance of 470 mi, the ratio was about 1.8 to 1. The higher ratio for the Big Muddy River is partly explained by the fact that section of river shown was specially selected for its display of a sequence of nicely formed meander loops.

Channel Changes in the Mississippi River Figure C is a composite map of a portion of the Mississippi River showing its channel in four surveys. The first, by Lieutenant Ross, was made in 1765; the last shown was made 135 years later. During this period the river was free to change its course without human intervention. Then, in the 1930s, a great series of artificial cutoffs was begun by the U.S. Army Corps of Engineers to shorten and straighten the river course.

(6) Study the Moss Island meander bend, which was cut off in 1821. How wide is the river channel here? Describe how the narrowing of the meander neck took place. In what part of the neck did most of the channel shifting take place, and in what direction?

_____

_____

_____

_____

(7) Study the growth of the two new meander bends indicated by the letters (a), (b), and (c) on the map. What event seems to have set off the growth of bend (a)?

_____

_____

_____

_____

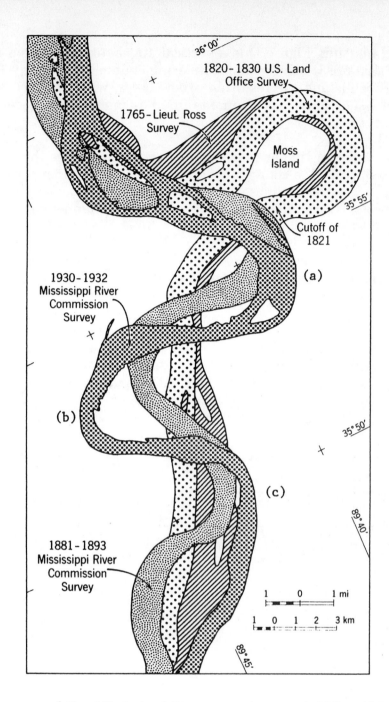

Figure C *Four surveys of the Mississippi River superimposed. (After U.S. Army Corps of Engineers.)*

(8) Downvalley motion of bends, accompanying their outward (sideward) growth is typical of alluvial river migrations; it is called *downvalley sweep*. What evidence of sweep can you find in the growth of bends (a), and (b)? About how much distance of sweep is indicated and in what period of time?

_____

_____

_____

Oxbow Lakes and Marshes Figure D is a vertical air photograph taken at an altitude of about 20,000 ft (6,100 m). It shows the meandering channel of the Hay River in Alberta, replete with cutoff bends and the resulting oxbow lakes, which in turn have been partly obliterated by new bends and their remnants filled by marsh (bog) vegetation.

(9) Fasten a sheet of tracing paper over the photograph and trace the map border. Using a color pen or pencil, trace the course of the present river channel across the photo. Using the same color, trace the outlines of what appear to be oxbow lakes (open water surfaces). In another color, trace all segments of older, filled channels, presumed to be bogs.

(10) Locate the most recent cutoff; label it as C-1. Locate a point at which the next cutoff is due to take place; label it as C-2. Find a likely candidate for the next cutoff and label it C-3.

(11) Study the surface features lying inside the narrowed bend in the lower right portion of the photo. What is the significance of the set of fine concentric curved lines lying within the bend? Trace some of these lines to establish their pattern.

_____

_____

_____

_____

_____

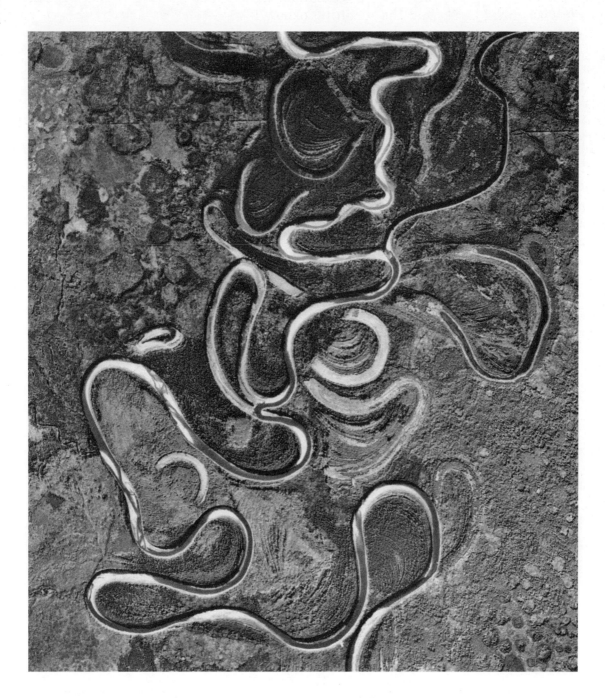

Figure D *Vertical air photograph of the Hay River, Alberta. (Lat. 58°55'N; long. 118°10'W.) (National Air Photo Library, Surveys and Mapping Branch, Canada Department of Energy, Mines and Resources. Photo No. A1-5183-38.)*

NAME _____ DATE _____

Exercise 16-D Alluvial Fans of Death Valley, California

[Text p. 536–537, Figures 16.29, 16.30.]

The typical North American landscape of mountainous deserts is perhaps best represented by the Death Valley region of southeastern California. Here, the natural landforms are protected from degradation by humans within the Death Valley National Monument. Figure A is a panoramic view northward into Death Valley. Dominating the foreground and middle distance of the scene are great alluvial fans, built of coarse alluvium derived from the bordering mountain. In this exercise, we look closely at a single fan in Death Valley.

The oblique color air photo, Figure 16.30 in your text, shows the general relationship of the Death Valley fans to the deeply dissected mountain range which feeds them with water and debris. It shows well the braided channels of the main fan-building streams and the myriad of smaller channels that are carving up the older fan surfaces.

For closer study, we refer to Figure B, another oblique air photo, but taken nearer to the mountain base. To assist you in finding specific features on the photo, we use a letter/number grid. The Panamint Range in the background rises to summit elevations of 10,000 to 11,000 ft (3,000 to 3,300 m). Surface elevations in the foreground are close to zero (sea level). Hanaupah Canyon is the main watershed on which the photo is centered.

Figure A *Air view of Death Valley, California. (Spence Air Photos.)*

Figure B *The Hanaupah Canyon alluvial fan in Death Valley, Inyo County, California. The view is west toward Telescope Peak in the Panamint Range. (Spence Air Photos.)*

(1) Follow the course of the main channel from its mouth at G-6 to E-3. Describe this channel and its boundaries.

_____

_____

_____

_____

(2) What conspicuous change does this main channel undergo downstream from about E-3? Describe what you see.

_____

_____

_____

_____

(3) Another conspicuous channel can be identified at J-3. Where does this channel originate? Why doesn't it join with main channel?

_____

_____

_____

_____

(4) Between the two channels referred to above there is a broad area of fan surface (vicinity of H-3) scarred by many narrow "wiggly" ravines. Where do these minor channels arise and what geomorphic activity are they performing?

_____

_____

_____

_____

(5) Where else on the fan are similar minor channels present?

_____

_____

_____

(6) What feature can you identify at F-6? Describe the activity taking place here.

_____

_____

_____

_____

(7) Summarize as best you can the geomorphic history of this fan as related to the features you have identified. What sequence of events can you postulate?

_____

_____

_____

_____

_____

_____

NAME _____   DATE _____

Exercise 17-A Mesas, Buttes, and Cliffs

[Text p. 546–547, Figures 17.4, 17.5, 17.6.]

An arid climate leaves much of the land surface barren of large plants. Consequently, we find that vertical cliffs, flat-topped mesas and plateaus, and isolated buttes stand out boldly, giving us some of the finest scenery in the American west. Torrential rains rapidly remove the soft, weak shales from under the edges of hard caprock layers, creating badlands that resemble mountains in miniature.

Monument Valley, found in the Navajo Country of Arizona and Utah, ranks close to first as a landscape of butte and mesas. Figure A shows the two "most photographed" buttes in the valley, if not in all the world. Called the Mitten Buttes—for obvious reasons—they are even recognizable as a matching left-hand and right-hand pair. That on the far right, Merrick Butte, is more conventional; it is one large rock mass. All three buttes are remnants of a single thick layer of massive sandstone that once extended over the entire area. Hollywood quickly discovered Monument Valley and capitalized on its beauty.

Figure A *Buttes of Monument Valley, northern Arizona. Mitten Buttes at left, Merrick Butte at right. (Infrared photograph by A.N. Strahler.)*

(1) Study the vertical sides of the three buttes in the photograph. What do you see that gives information on the process of cliff retreat and butte formation?

_____

_____

_____

_____

_____

Figure B is a contour topographic map of mesas and buttes in the central part of Monument Valley. Cliffs are indicated by the bands of closely crowded contour lines. Keep in mind that, if vertical cliff faces are correctly represented, contours on the cliff face will be "stacked" one upon the other to make a single line.

(2) Locate as exactly as possible the point on the map where the photograph, Figure A, was taken (along the road to Lookout Point). Draw rays from the point to indicate the area encompassed in the photo.

(3) About how high do these buttes rise above the level of the surrounding lowland?

Summit elevation: _____ ft. Contour near base: _____ ft.

Difference: _____ ft.

(4) The height of the vertical sides (cliffs) of these buttes is difficult to determine from the map contours. Judging from the photo the proportion of the height (Question 3) that can be assigned to the cliff, how thick is the massive sandstone formation?

_____

_____

(5) Judging from its appearance in the photo, what kind or kinds of sedimentary rock make up the pedestals of the buttes? Give reasons for your answer.

_____

_____

_____

_____

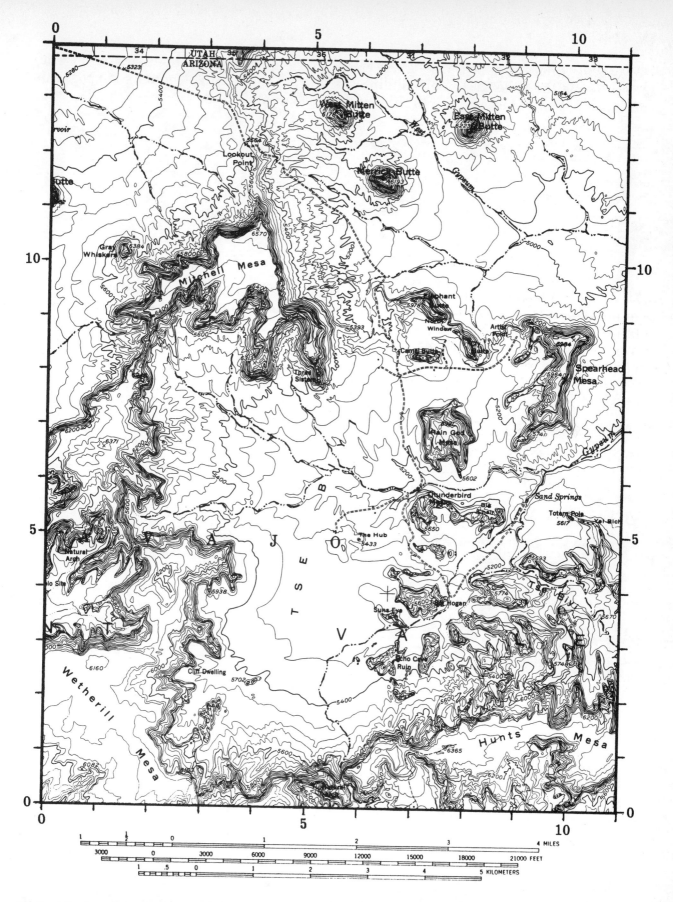

Figure B *Portion of the Agathla Peak, AZ-UT, Quadrangle. Scale 1:62,500. Contour interval 40 ft. (U.S. Geological Survey.)*

(6) Judging from the appearance of the slopes in the foreground of the photo, what kind of rock forms the valley floor surrounding the buttes? Do the map contours support this conclusion? What name is given to this kind of topography?

_____

_____

_____

_____

(7) Three mesas conspicuous on the map for their broad, flat tops, are named Mitchell, Wetherill, and Hunts. Read and record the elevation of a contour near the middle of each of these mesas. Assuming that these elevations signify the upper surface of the sandstone formation, what can you conclude about the attitude of that rock layer? Is it horizontal or does it slope (dip)? If the latter, in what direction?

_____

_____

_____

_____

_____

NAME _____   DATE _____

Exercise 17-B Domes and Their Hogbacks

[Text p. 551–553, Figures 17.9, 17.10, 17.11, 17.12.]

Sedimentary domes are not common geologic structures, but some are found in the Middle Rockies and bordering parts of the Great Plains. Nearly circular domes are rare in this region. More commonly they are elliptical.

Before proceeding further, we should think about how circular and elliptical sedimentary domes are formed and what lies beneath them. Two quite different origins are recognized, with excellent evidence for both. One is the *laccolithic dome* formed by an igneous intrusion between sedimentary formations. It is a relative of the igneous sill, described in Chapter 11, and is illustrated in our Figure A. Instead of spreading widely, the magma pushes up the overlying strata into a circular dome. The igneous mass itself is called a *laccolith*. An excellent example of a laccolithic dome is Navajo Mountain, rising from the Rainbow Plateau in southern Arizona. It lies close to the famed Rainbow Bridge near Lake Powell.

Larger sedimentary domes, such as the Black Hills Dome (text Figure 17.12), represent uplifts of the crust deep beneath the surface, and they are interpreted as tectonic in origin. Figure A compares tectonic and laccolithic domes. The lower diagram shows that ancient crustal rock, Precambrian in age, lies under the strata and has been elevated along with the strata. Recent seismic studies show that several such domes and arches of the central Rocky Mountain region were pushed up by overthrust faults that rise steeply under the dome. In some cases, the thrust plane cuts through to the surface. This kind of thrusting would explain domes with steep dips on the eastern side, but gentle dips on the western.

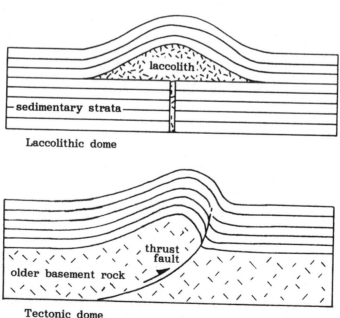

Figure A *Comparison of a tectonic dome with a laccolithic dome. (A.N. Strahler.)*

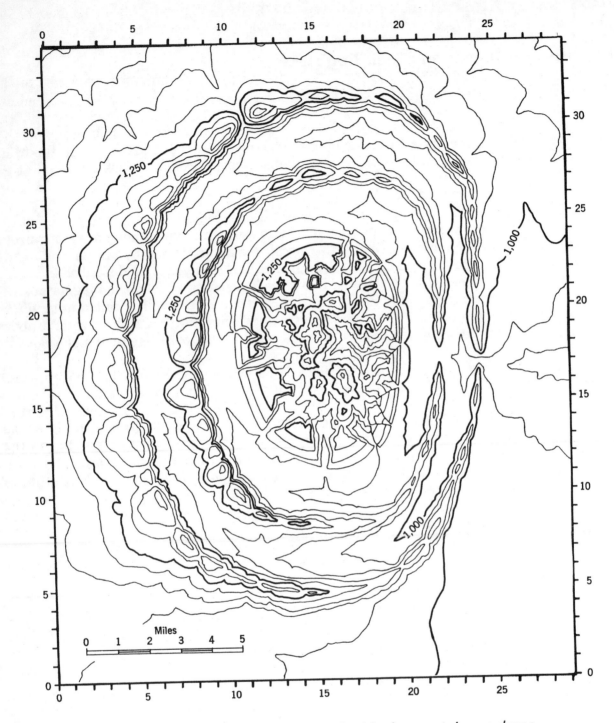

Figure B Synthetic topographic contour map of a ideal mountainous dome.
(A.N. Strahler.)

Figure B is a topographic contour map of an imaginary sedimentary dome of the tectonic variety. It's a sort of cross between the circular dome diagrammed in text Figure 17.9 and the large elongate Black Hills dome of Figure 17.12. Given the help of these two text figures, you should be able to interpret the contour map without difficulty.

(1) Hogbacks are described in your text and illustrated in Figures 17.9, 17.10, and 17.12. Find two good examples on the topographic map and label them with the letter H.

(2) Flatirons are not mentioned in your text; they are special landforms related to hogbacks. Where a steeply upturned hard formation is cut through by closely spaced streams, the hard layer is shaped into a triangular slab, point-up, as in the sketch below. Find flatirons in diagram *b* of text Figure 17.9. Locate two good examples of flatirons on the our topographic map and label them with the letter F.

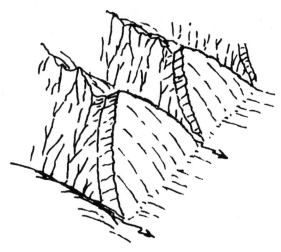

Figure C ties together the landforms of horizontal strata, gently-dipping strata (such as those of coastal plains), and domes. One group of landforms grades into the other on the diagram, as the dip changes from zero at the rear to 45° or more on the front face of the diagram. Subsequent streams (S) occupy long, narrow valleys of weak shale. These are separated by cuestas (Cu) and hogbacks (H).

(3) On the topographic map (Figure B) locate two subsequent valleys and label them SV. Draw the subsequent streams in these valleys, using a red pen or crayon; label them S.

(4) Examine the central crystalline area of the dome as shown by contours on the map; label it CCA. What drainage pattern is developed here? If this were an arid region, what distinctive landforms might you expect to find here? Draw in black line the stream channels of the central crystalline area.

———

———

———

(5) Construct a topographic profile from 0-9 to 29-19 across the topographic map, using the blank graph provided (Graph A). Add to the profile a geologic structure section consisting of layers of sandstone and shale conforming to the hogbacks, flatirons, and subsequent valleys. Show crystalline rocks (granite pattern) in the core of the dome. (Using clear tape, replace the completed profile in its original place on the page.)

(6) Assuming that the sandstone formations that make the two hogback ridges on the map (Figure B) are of uniform thickness over this entire region, explain why the ridges are broad and cuestalike on the west side of the dome, but narrow and straight-crested on the east side of the dome. Does your answer explain why the major streams drain out through the east side of the dome?

_____

_____

_____

_____

_____

_____

_____

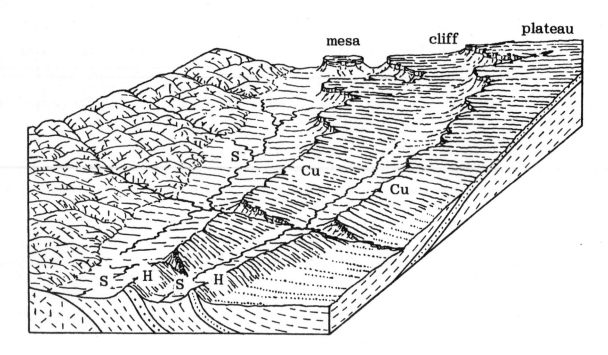

Figure C _Schematic diagram of landforms transitional from horizontal strata to steeply-dipping strata. (After W.M. Davis.)_

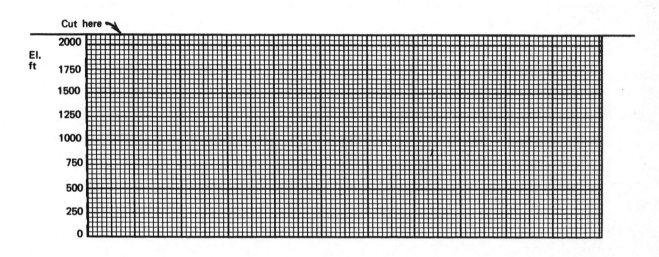

Cut here

El.
ft

2000
1750
1500
1250
1000
750
500
250
0

Graph A

NAME _____ DATE _____

Exercise 17-C Mountain Ridges on Folded Strata

[Text p. 533, 548–549, Figures 17.13, 17.14, 17.15.]

Foreland folds were introduced in Chapter 13, illustrated by the Zagros Mountains of Iran and the Jura Mountains of the northern European Alps. Keep in mind that in both examples the folding is geologically very recent—Late Cenozoic—and resulted from the collision between the Africa plate and the Eurasian plate. In text Figure 17.13, block *a* illustrates the Zagros-type folds, in which most of the mountains are anticlines. They are in the class of initial landforms—landforms produced directly by tectonic or volcanic processes. We now turn to mountains developed by deep erosion (denudation) of very ancient fold structures; they belong to the class of sequential landforms.

Deeply eroded anticlines and synclines form striking zigzag ridges in the Ridge and Valley Province (or Folded Appalachians), a narrow belt that extends from Pennsylvania southward through Maryland and Virginia and ends in Alabama. Here, strata of Paleozoic age were thrown into foreland folds in the continental collision that took place in late Carboniferous (Permian) time, about 240 million years ago. Since then, the regional history has been largely one of denudation, exposing deeper and deeper parts of the fold structures. Today's topography is illustrated in text Figure 17.13, block *b*. Note especially that synclines as well as anticlines form the high ridges, and some of the valleys lie along the axes of anticlines.

Zigzag Ridges on Folded Strata Examine closely the side-looking airborne radar (SLAR) image of zigzag ridges of the Ridge and Valley Province near Hollidaysburg, Pennsylvania, text Appendix II, Figure A2.1. Find this area on the Landsat image, text page 548. Note that the location of the SLAR image is along the northwestern edge of the fold belt and includes a small portion of the Allegheny Plateau (upper-left corner), where the strata are nearly horizontal. Your objective now is to learn more about the special landforms found in an area of zigzag ridges developed on plunging folds.

Figure A reproduces text Figure 17.15 with added lines and labels. A dashed line is drawn along the axis of the syncline and another along the axis of the anticline. Arrowheads at the far ends of these two lines show that the direction of *plunge* of the folds is away from the observer. Study the mountain that ends in a curved cliff at the point of crossing the synclinal axis; this feature is a *synclinal nose*. Following the same cliff toward the right to the point where it reaches the anticlinal axis, we observe that it doubles back on itself, enclosing an *anticlinal cove*. Away from this summit the mountain slopes gently down to a point and disappears; this feature is an *anticlinal nose*.

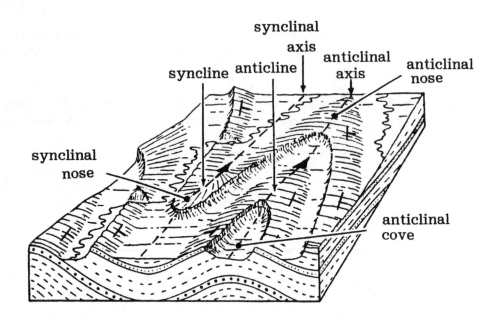

Figure A *Landforms on eroded plunging folds. (After E. Raisz.)*

(1) Describe the two kinds of "noses" and explain how their contrasting topographic forms are related to geology.

_____

_____

_____

_____

_____

_____

Plunging Folds Shown by Topographic Contours Figure B is a synthetic topographic map, not representing any real area, but designed to illustrate the typical landforms of the Appalachian Ridge and Valley Province. For guidance in interpreting the map, Figure C is an idealized geological cross section that shows both the landforms and structures of a series of deeply eroded open folds. A sandstone formation (ss) and a conglomerate formation (cgl) are the ridge-formers; the thick shale formations above and below are the valley-formers.

Below this diagram are listed three varieties of ridges (mountains) and three varieties of valleys. Of these, four are features labeled and illustrated in text Figure 17.13. The other two are new terms, not found in your textbook. A ridge in which the dip of the hard layer is in one direction only is called a *homoclinal ridge* (HR). It is identical in definition and form to the hogback, a ridge developed in a deeply eroded dome (Exercise 17-B). A valley in which the strata dip in one direction only is a *homoclinal valley* (HV). Thus we now have a set of six terms in all to describe the ridges and valleys.

(2) Identify and label each ridge and valley on the map, using the initials given in the key to Figure C. Enter numerous strike-and-dip symbols on the ridges. Show the direction of plunge of fold axes by long arrows. Label the arrows A for anticline; S for syncline.

(3) Draw lines to show a complete drainage system for the map area. Extend the streams shown on the map as needed to complete the minor branches. Use the color red for all subsequent streams (S).

(4) What name is given to the kind of drainage pattern shown on this map? (See text Figure 17.14.) Describe the pattern in terms of the types of streams that compose it. Compare it with the drainage pattern of a deeply eroded dome (text Figure 17.11).

_____

_____

_____

_____

_____

(5) (Optional) Construct a topographic profile from the upper left-hand corner of the map to the lower right-hand corner. Use the blank profile graph provided (Graph A). Draw in a complete geologic structure section consisting of sandstone and shale formations in agreement with your map interpretations already made and labeled. After completion, return the profile to its former position on the page and secure it with tape.

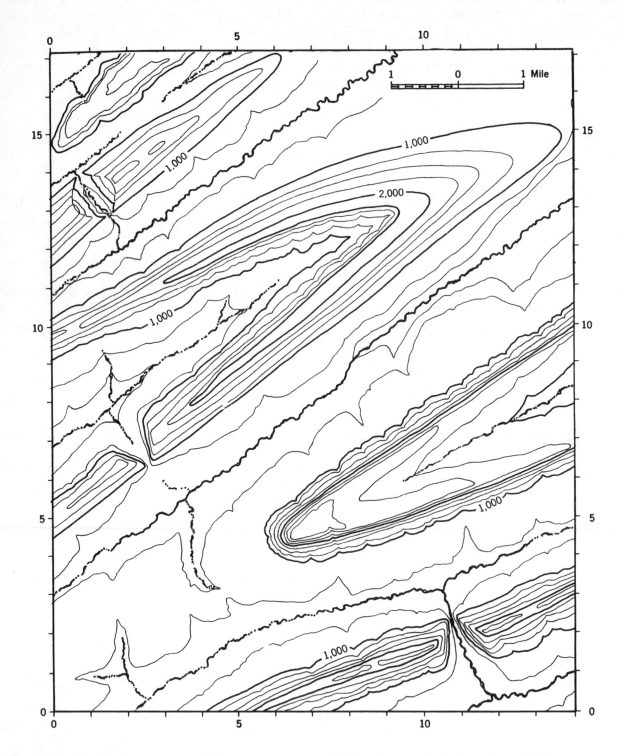

Figure B *Synthetic topographic contour map of deeply eroded plunging folds.*
(A.N. Strahler.)

Copyright © by Arthur N. Strahler

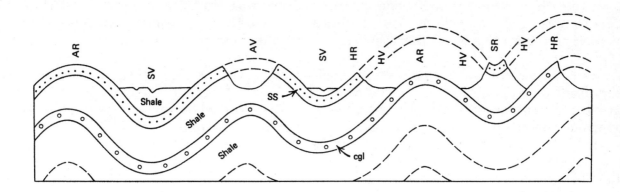

AR   Anticlinal ridge or mountain   AV   Anticlinal valley

SR   Synclinal ridge or mountain   SV   Synclinal valley

HR   Homoclinal ridge or mountain   HV   Homoclinal valley

Figure C Schematic cross section of relationship of ridges to fold structures. (A.N. Strahler.)

Watergaps and Windgaps Throughout the Appalachian Ridge and Valley Province we find numerous examples of streams that cross from one subsequent valley to another by passing through a deep, narrow watergap in the intervening ridge. On the Landsat image, text Figure page 548, several major watergaps of the Susquehanna River are clearly shown in the upper right-hand corner.

(6) On the topographic map, Figure B, locate two watergaps. Give grid coordinates for each and identify the kind of ridge through which each is cut. Describe in detail the topographic form of each gap.

_____

_____

_____

_____

_____

_____

Stream Capture Throughout the Appalachian ridges we occasionally find what looks like a watergap in a ridge, but on closer look has no through-going stream occupying the gap. A still closer look may show that the floor of the gap is considerably higher in elevation than the valley floor on either side. Such a feature was named a *windgap* by early settlers to the region. They often took advantage of a windgap to accommodate an easy trail or road across a ridge that otherwise was a formidable barrier to travel.

Windgaps are correctly interpreted as former watergaps in those cases where no fault zone is present to explain the gap. As diagrammed in Figure D, the abandonment of a gap by the stream that formerly occupied it and carved it is explained by *stream capture*, a process by which the headwater drainage basin of one stream, the *captor stream*, is gradually extended headward toward the trunk of another stream that is situated at a higher elevation. Eventually, the flow of the higher trunk stream is diverted into the favored captor, and this leaves as its victim a *beheaded stream* with only a trickle of flow in its former channel below the point of capture (Figure D, "After"). Our diagrams show that the captor stream is a tributary of a large river, with a profile at low elevation and capable of easily eroding its watergaps to keep its profile low. The victim, a small stream, has difficulty maintaining its course across the hard rock barrier of its watergap; its elevation remains high. It easily succumbs to the predator stream, which has only a short distance to flow on weak rock to meet the main river. The key to this process is difference in elevation of the two streams in the divide area that separates them.

(7) Our imaginary topographic map, Figure B, shows a windgap at 2-7. Using a special color—such as a yellow or pink highlighter—show the streams as they were before the capture. Which stream was the captor? Why was it successful? State your evidence.

_____

_____

_____

_____

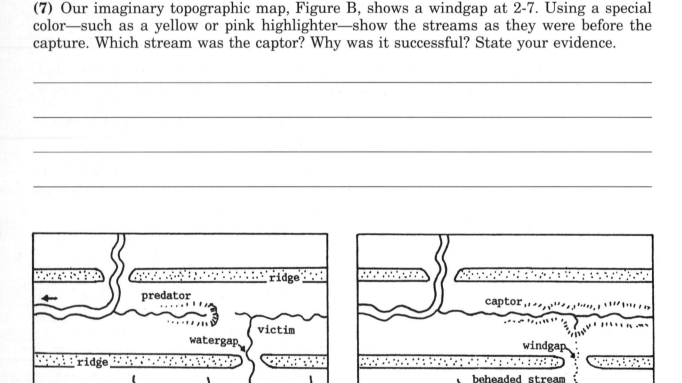

Figure D *Schematic maps of stream capture, leaving a windgap. (A.N. Strahler.)*

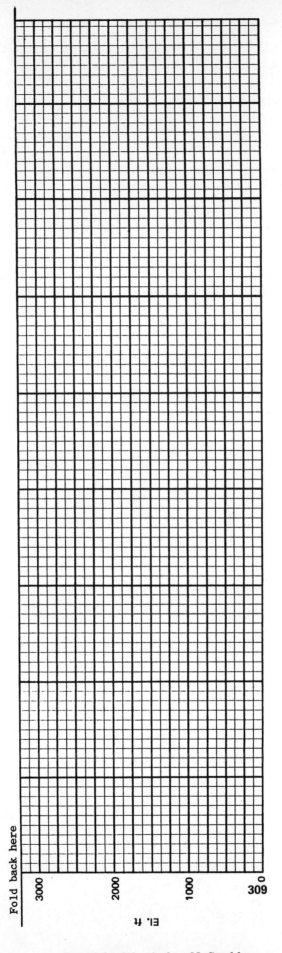

Graph A

NAME _____ DATE _____

Exercise 17-D Stone Mountain—A Granite Monadnock

[Text p. 555–556, Figure 17.19, 17.20.]

Granite exposed in domes seems to excite and attract sculptors, especially those who work on a grand scale, using pneumatic jack hammers and explosives. In the granite of Mount Rushmore of the Black Hills uplift, Gutzon Borglum carved those great busts of four American presidents, visible from as far away as 60 mi (100 km). Borglum died in 1941, and the work was completed by his son, Lincoln. Years earlier, in Georgia, the Daughters of the Confederacy had commissioned the carving of equestrian statues of Robert E. Lee, Stonewall Jackson, and Jefferson Davis in the steep granite face of Stone Mountain, near Atlanta. Borglum designed and partially completed the work, but in 1924 resigned in a tiff with his sponsors and destroyed his plans. Two other sculptors worked on it, and despite long delays, it was dedicated in 1970. The State of Georgia now owns Stone Mountain and has transformed it into a state park.

As your text explains, Stone mountain is a monolith of granite, what remains of an ancient granite pluton intruded into metamorphic rocks of the Piedmont region. Throughout most of the Cenozoic Era, this region has been undergoing denudation, and most of it has been reduced to a rolling upland, called the Piedmont Peneplain. Numerous monadnocks rise above that level; many of them are of quartzite and have elongate, ridgelike forms. Stone Mountain is perhaps unique in the Piedmont in terms of its compact outline and smoothly rounded form, but there's a similar one in the Blue Ridge near Asheville, North Carolina. Other prominent domes compete for attention in other lands. Rio de Janiero has its granite Sugar Loaf, while Australian tourism extolls Ayres Rock, a prominent sandstone mass in the desert of Northern Territory. But Pasadena, California, has its own little-league contender in Eagle Rock, a small but prominent dome of massive conglomerate. Actually, Australia has some fine granite domes resembling Stone Mountain, but they are on the Eyre Peninsula of South Australia. Splendid granite domes also occur in the Nubian Desert of North Africa. And what about the Yosemite Domes? We studied them in Exercise 13-A as exfoliation domes.

We investigate Stone Mountain with the help of a photograph and two topographic contour maps. The old oblique air photograph, Figure A, shows the steep northeast face of the mountain, hidden from view in the color view of text Figure on page 559. Details of the dome are shown on the large-scale map, Figure B, while the relationship of the dome to the surrounding Piedmont upland is better seen in the small-scale map, Figure C.

Figure A *Taken in the 1920 by the U.S. Army Air Corps, this photo of Stone Mountain shows the scenery long before development of the area as state park.*

(1) In what important way or ways does Stone Mountain differ geologically from the Yosemite domes of Exercise 14-A?

_____

_____

_____

_____

_____

_____

(2) Locate on the map, Figure B, the ground point above which each of the two photos was taken. From each point, draw an arrow across the dome to show the direction of view. Label your arrows P. 559 and Fig. A.

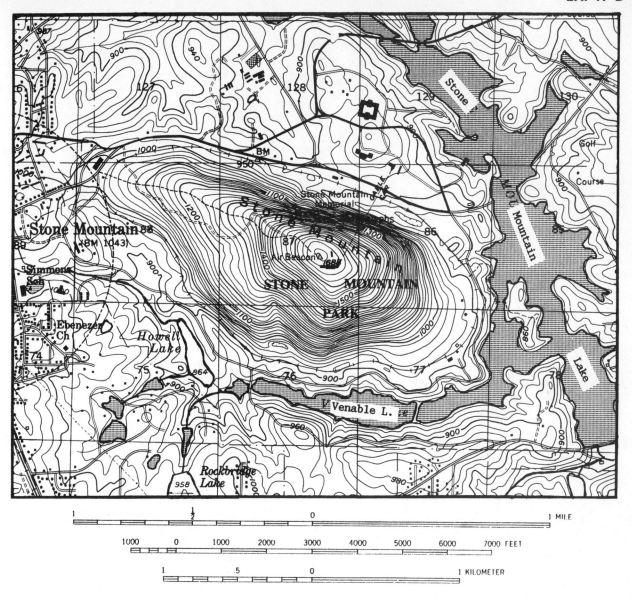

Figure B Portion of the Stone Mountain, Georgia, Quadrangle. U.S. Geological Survey, 1954. Scale 1:24,000. Contour interval 20 ft.

(3) What is the nature and cause of the parallel dark streaks or lines on the flanks of Stone Mountain?

_____

_____

_____

_____

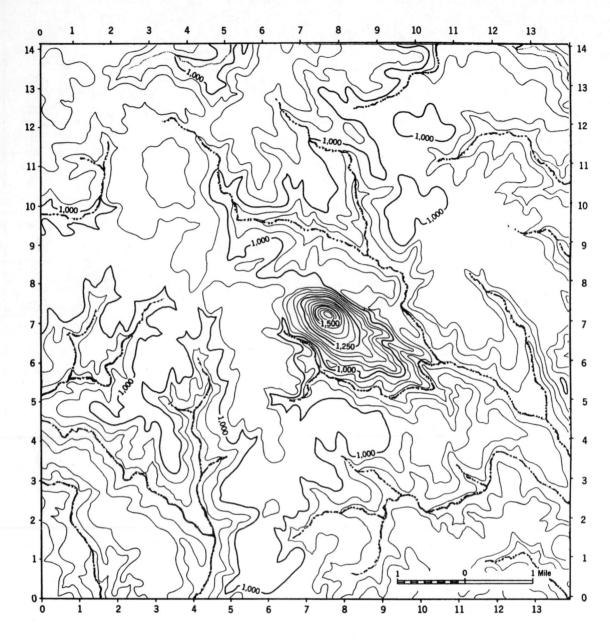

Figure C *Portion of the Atlanta, Georgia, Quadrangle, U.S. Geological Survey. Scale 1:125,000. Contour interval 50 ft. (Map scale has been enlarged to about 1:87,000.)*

(4) Using the small-scale topographic map, Figure C, construct a topographic profile through the summit of Stone Mountain from 0.5-1.5 to 12-11. Use the blank graph (Graph A). When the profile is completed, replace it and secure with tape.

(a) Below the profile, draw the outlines of the resistant granite stock that lies beneath Stone Mountain. Label it **granite**.

(b) Using a color pen or pencil, draw a horizontal line across the profile at the approximate level of the Piedmont Peneplain, which can be taken as between 1000 and 1050 ft.

The Piedmont Peneplain reached its full development many millions of years ago, and was then raised during a broad crustal uplift that affected the entire eastern part of North America. Streams on the peneplain became rejuvenated and trenched their valleys below the peneplain level, which is now represented only in the higher summits of the area surrounding Stone Mountain.

(5) Using blue pencil or pen, draw a complete drainage pattern on map of Figure C. Show streams wherever indicated by a V-indentation of the contour lines. What kind of drainage pattern is shown? Is it similar to that shown in text figures 17.6 and 17.19? What general statement can you derive from your comparison?

_____

_____

_____

_____

_____

_____

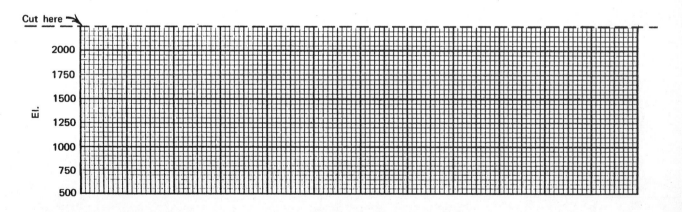

Graph A

NAME _____   DATE _____

Exercise 17-E A Volcanic Neck and Its Radial Dikes

[Text p. 556–560, Figures 17.21, 17.24.]

In an arid climate, volcanic necks that were one surrounded by thick shale formations now rise as striking isolated peaks, surrounded by plains. They are locally called "buttes," but there's a world of difference between their geology and that of the sandstone buttes, such as the Mitten Buttes of Monument Valley (Exercise 17-A). While there are quite a few volcanic necks in the Colorado Plateau region of northeastern Arizona and northwestern New Mexico, Ship Rock certainly reigns supreme.

Figure A shows comparison cross-sections of a volcanic neck, such as Ship Rock, and a sedimentary butte, such as the Mitten Buttes of Monument Valley. Notice that the massive igneous rock of the volcanic neck has a set of small, irregular vertical dikes. Rough vertical joints are also present, and weathering develops sharp points of varying heights.

We investigate Ship Rock and its radial dikes by means of two photographs and a contour topographic map. Figure B is an oblique air photo taken many years ago by Robert E. Spence, a noted documentary photographer of western scenes, mostly in southern California. The color photograph in your textbook, Figure 17.24, shows Ship Rock from a different angle.

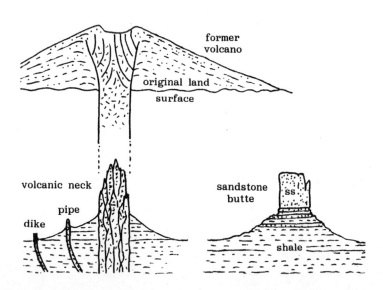

Figure A *Sketch comparing a volcanic neck with a sandstone butte. (A.N. Strahler.)*

Figure B *Oblique air photograph of Ship Rock, New Mexico, and its radial dikes. (Robert E. Spence.)*

(1) On the topographic map, Figure C, find the points above which the plane was located when each of the two photos was taken. Draw a line from each to show the center line of the photo. Label the lines Fig. 17.24 and Fig. A.

(2) Determine the contour interval used on the map. Estimate the summit elevation of Ship Rock and its height above the surrounding plain. How high is the dike at 6.7-1.2?

Contour interval: _____ ft.

Summit elevation: _____ ft, approx.

Height above plain: _____ minus _____ equals _____ ft.

Height of dike: _____ ft.

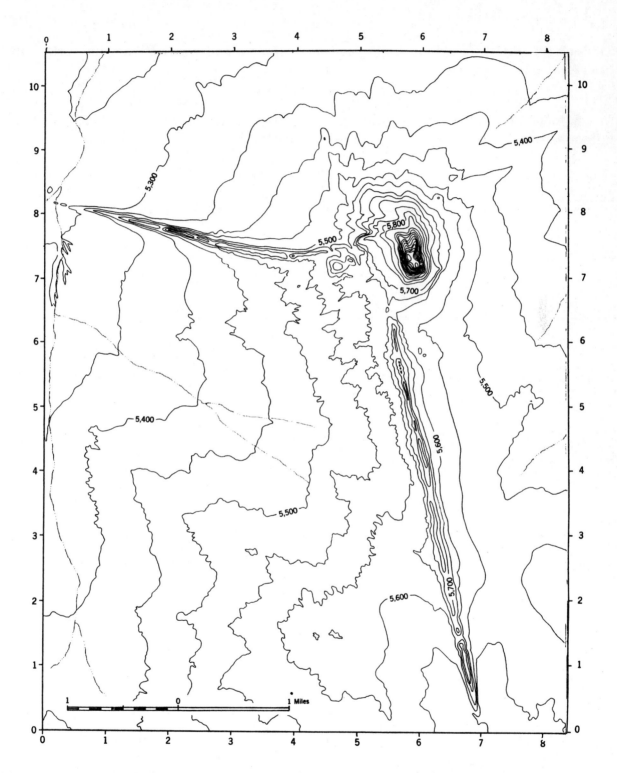

Figure C *Portion of Ship Rock, New Mexico, Quadrangle, U.S. Geological Survey. Scale 1:62,500.*

(3) On the map, color in red all exposed volcanic rock. The dikes should be shown as thin red lines. In addition to the two great dikes, at least two small dikes are indicated by sharply pointed contour lines. Label the volcanic neck and the dikes.

(4) With the aid of the photo, Figure B, locate three small pipes of volcanic rock. Label these and color them red.

(5) Contours near 5.0-6.0 are highly crenulated (crinkled). What is the meaning of this contour form? What landform type is present. What kind of rock is indicated?

_____

_____

_____

_____

_____

Exercise 18-A The Ria Coast of Brittany—A Youthful Shoreline

[Text p. 572–574, Figure 18.13.]

The ria coast, listed first in your textbook under coastlines of submergence, gets it name from the northeastern coast of Spain, where numerous capes project out into the Atlantic and between them lie narrow, branching bays. The title *Ria* is used for these bays—Ria de Betanzos and Ria de Muros y Noya, for example. In Spanish *ria* means "estuary."

A ria coast must meet certain requirements. First, it is a partially drowned (submerged) land surface made up of erosional landforms. Tectonic and volcanic landforms are excluded. This means we are dealing with land surfaces eroded by (a) fluvial processes (streams, along with mass wasting, and weathering) or (b) glaciers, either alpine or continental. For ria coasts, the choice is (a), so we expect to find river valleys in a partially drowned condition. A close look at the bottom topography of the bays is necessary in order to identify valleys in which streams formerly flowed.

Our exercise is based on a topographic contour map (Figure A) of the coastal region around the port of Brest, France. It lies on the peninsula of Brittany, which projects westward into the Atlantic Ocean. Similar in many ways to the Ria coast of Spain, the Brittany coast is deeply embayed, and it is modified strongly by wave erosion only where the shore is exposed to waves of the open Atlantic Ocean. The coast was never modified by Pleistocene glaciers or ice sheets.

The contour interval of the map is 20 m on land. Submarine contours are shown as dashed lines for depths of 1, 5, 10, 20, 40, and 50 m. A stippled pattern shows areas that lie between the high-water line (bold line) and the low-water line (fine line). Marine cliffs of rock are shown by lumpy (bumpy) projections along the high-water line.

(1) In red pencil or pen, mark on the map all parts of the shoreline where a marine cliff is well developed. In green, shade all probable sand or shingle beaches. ("Shingle" means well rounded pebbles or cobbles.) Label three examples of beaches; three of marine cliffs. Find and label a good example of a pocket beach on the outer shoreline.

(2) Using a pencil, draw lines on the map to give a reconstruction of the drowned stream system in the Brest harbor, or estuary. Reconstruct branches that connect with the mouths of small streams that enter side branches of the estuary. Carry the lines inland, up the tributary valleys.

(3) Locate two meander bends of the former stream system. (Give grid coordinates.)

  Meanders: _____ and _____

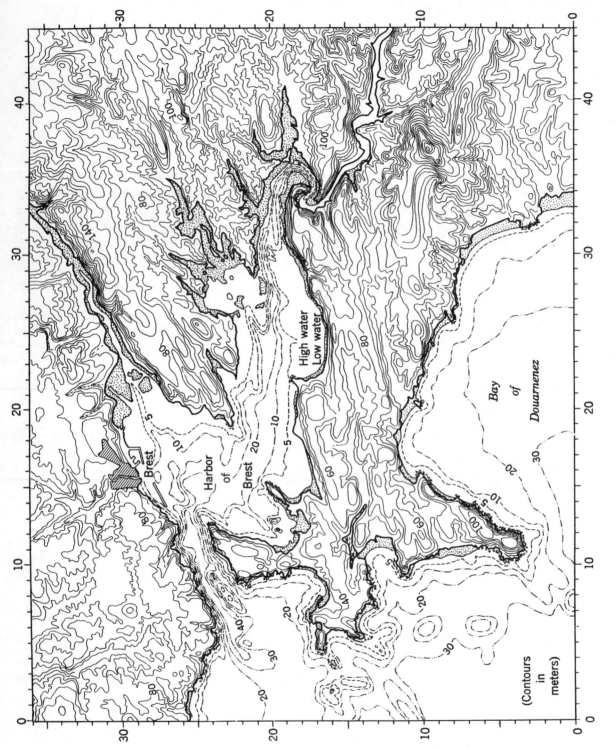

Figure A Portion of the Brest, France, topographic map, Sheet 21. Original scale 1:200,000.(4) Why

(Contours in meters)

(4) Why are there few prominent cliffs along the shoreline of the Harbor of Brest? Give the grid coordinates of cliffs that are indicated on the map within the harbor.

_____

_____

_____

_____

(5) Study the peninsula ending at 11-4. Which side seems to have undergone the greater amount of marine erosion? What is the topographic evidence? Is this configuration what you would expect, knowing that the open Atlantic Ocean lies to the west, whereas a bay 15 km wide lies to the east?

_____

_____

_____

_____

_____

_____

_____

(6) The stippled zone between high and low water levels is alternately exposed and inundated as the tide falls and rises. In the heads of bays near 30-20, what type of sediment deposit lies in the stippled zone?

_____

_____

_____

NAME _____ DATE _____

Exercise 18-B Coastal Sand Bars

[Text p. 569–571, Figures 18.5, 18.6, 18.7.]

As a ria coast is attacked by storm waves, promontories and headlands are cut back rapidly. At first there is little excess sand to form beaches, but at a later stage beaches appear in a wide variety of forms. Chances are that you have visited some of these kinds of beaches without knowing how they got there.

Depositional Forms of a Ria Coastline Your textbook doesn't describe these depositional features, so we offer an illustration (Figure A) showing them as produced during the evolution of a ria coastline. Much the same depositional forms are typical of embayed glaciated coasts, such as those of New England and the Maritime Provinces.

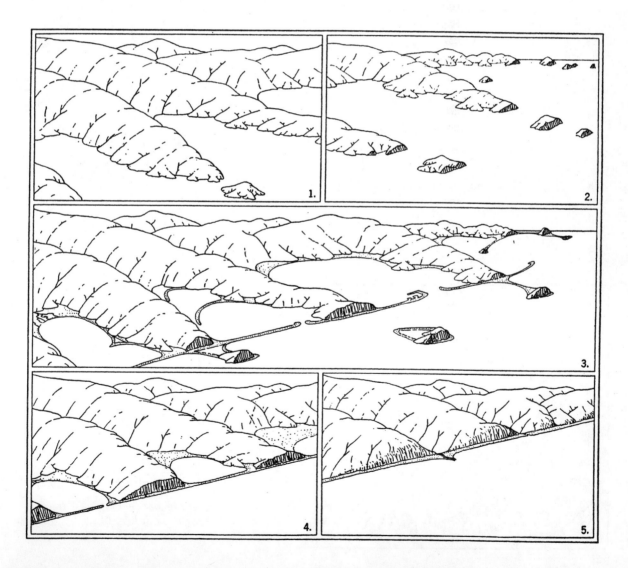

Figure A *Stages in the evolution of a ria coastline. (A.N. Strahler.)*

EX. 18-B

(1) Using the list below, enter the code letter of each landform at the place it is shown on the diagram. Because these landforms aren't defined in your textbook, you will need to make guesses. To help with the identification we offer sketch maps of a few of them, in turn-of-the-century cartographic styling (Figure B).

T Tombolo	CH Cliffed headland	BHB Bayhead beach
S Spit	DT Double tombolo	BSB Bayside beach
RS Recurved spit	HB Headland beach	BHD Bayhead delta
CS Complex spit	BMB Baymouth bar	L Lagoon
CT Complex tombolo	MBB Midbay bar	I Inlet
LB Looped bar	CB Cuspate bar	CP Cuspate delta

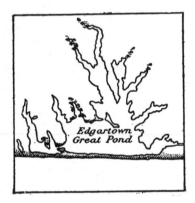

Baymouth bar sealing off a freshwater pond on Marthas Vineyard, MA.

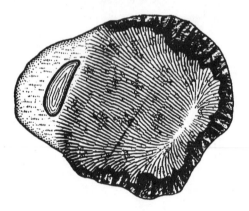

Looped bar on the lee side of Shapka Island, AK.

Bayhead bar near Duluth, MN.

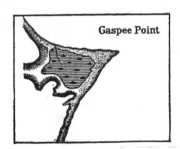

Cuspate bar, enclosing a marsh, near Providence, RI.

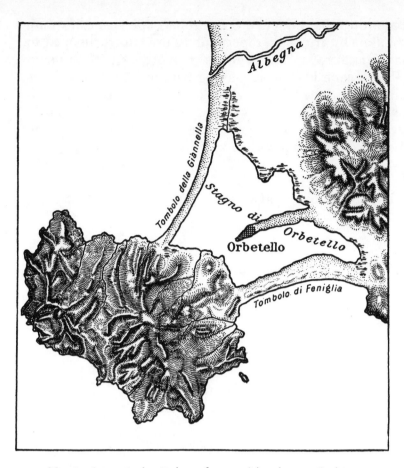

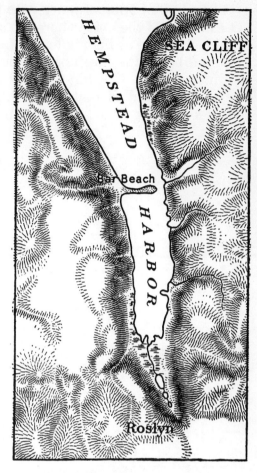

Monte Argentario, Italy, a former island now tied to the mainland by a double tombolo. The town of Orbetello is on an earlier, incompletely formed tombolo.

Bar Beach in Hempstead Harbor, Long Island, NY, is a midbay bar.

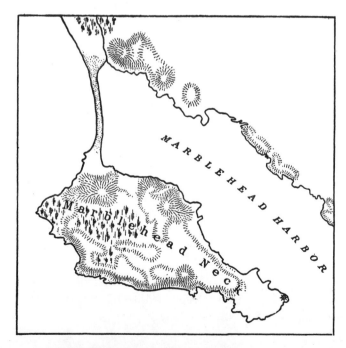

Marblehead Neck, MA, a former island connected to the mainland by a single tombolo.

Figure B Seven sketch-maps showing a variety of depositional coastal landforms. (From D.W. Johnson, Shore Processes and Shoreline Development. Copyright © 1919 by Douglas W. Johnson. Reprinted by permission of John Wiley & Sons, New York.)

A Shoreline of Submergence with Baymouth Bars Figure C is a topographic contour map of a section of the north shoreline of Lake Ontario. The topography here was developed in preglacial time by fluvial denudation that carved stream valleys. The area was then heavily scoured by the Erie Lobe of the Pleistocene ice sheet, leaving behind glacial till. Lake sediments were added in a late glacial stage, when Lake Ontario was larger than it is today. Waves and currents, reworking and shifting the cover of glacial and lake deposits have formed two sand bars that now separate Yeo Lake and Spence Lake from the open water. Sandbanks Provincial Park now occupies the coastal zone here.

(2) Which of the two sand bars is a baymouth bar, and which a midbay bar? Label them accordingly.

(3) At 12.5-6.0, the sand bar has long narrow ridges, parallel with the shoreline. Near 7-8 the sand bar has several very small hill summits. How do you interpret these features? What caused them?

_____

_____

_____

_____

(4) Describe the form of the narrow outlet channel through the bar between Spence Lake and Athol Bay. Explain in terms of beach drifting processes. Are tidal currents responsible for keeping this channel clear? If not, what keeps the channel clear?

_____

_____

_____

_____

(5) Using a color pencil or pen, redraw the shoreline as a smooth, simple shoreline of the future, after the headlands have been cut back to a line passing through 0.0-14.0, 6.0-11.0, 10.0-8.0, and 19.0-0.0. Redraw the sand bars in their new positions.

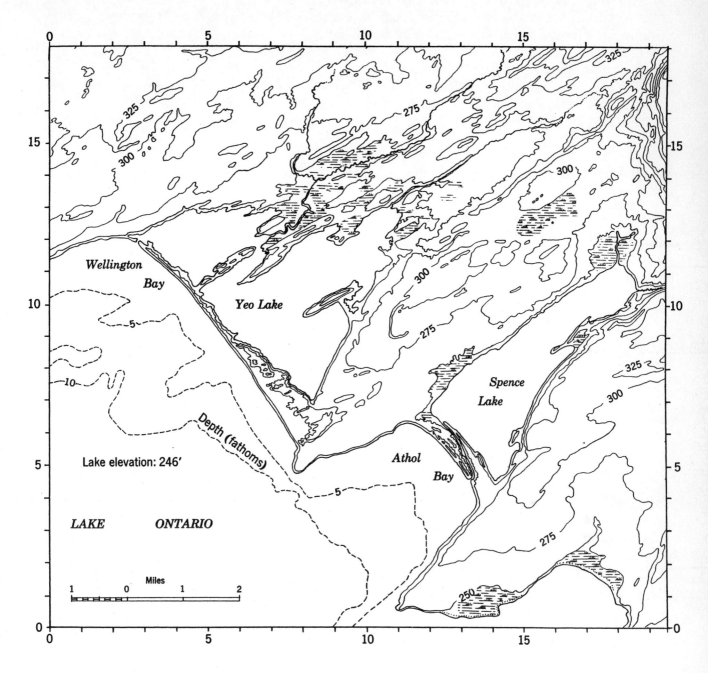

Figure C *Portion of Wellington, Ontario, topographic sheet. Geological Survey of Canada. Original scale 1:63,360.*

NAME _____ DATE _____

Exercise 18-C Barrier Island Coasts

[Text p. 571–575, Figures 18.13, 18.14, 18.15.]

Barrier island coasts are well represented along the passive eastern continental margin of North America, with its partly submerged coastal plain extending from New Jersey to Georgia, and around the entire Gulf of Mexico from the Florida panhandle to Tampico, Mexico.

Barrier Islands of the Delmarva Peninsula Figure A is a topographic contour map of a section of the barrier island coast in Virginia. The area shown is in the narrow southern end of the Delmarva Peninsula that lies between Chesapeake Bay and the open Atlantic and ends in Cape Charles. Not far to the north along this coast is Assateague National Seashore and the community of Chincoteague, familiar to many TV viewers as the locality where a herd of wild horses has its home.

Metomkin Island and Cedar Island are two segments of the barrier island, separated by Metomkin Inlet. Large areas of open lagoon—Metomkin Bay and Floyds Bay—compete with salt marsh, shown in a distinctive pattern laced by a network of sinuous tidal creeks. The mainland can be taken to begin along a line coinciding with the 10-ft contour, which defines the outer limit of flat upland surfaces bearing the local designation of "necks."

(1) Note that the four necks—Parker, Bailey, Joynes, and Custis—are sharply defined on the east by the 10-foot contour. This suggests the presence of a more-or-less straight scarp sloping down to tidewater level. The landward limit of the salt marsh is a nearly straight line paralleling that contour. Can you suggest an origin and history for this scarp?

_____

_____

_____

_____

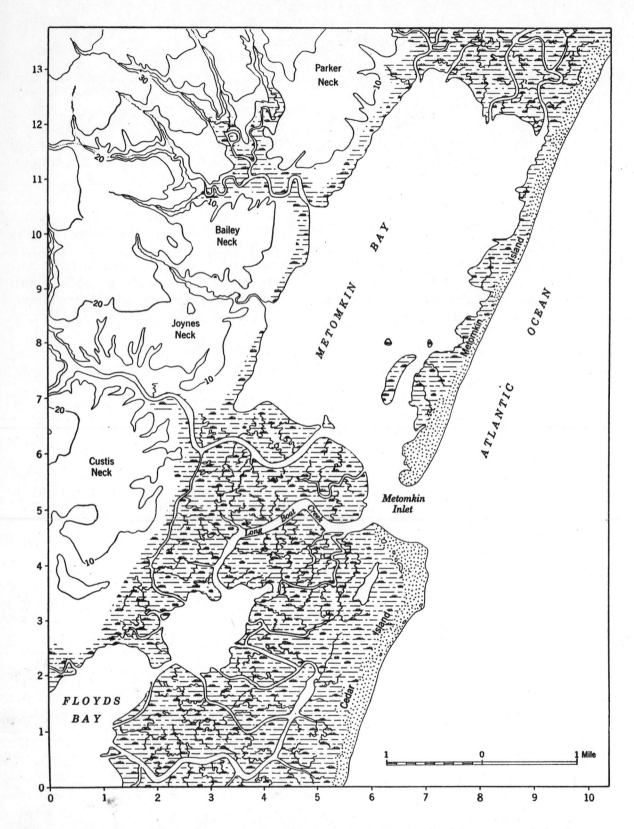

Figure A *Portion of Accomac, VA, Quadrangle. U.S. Geological Survey. Original scale: 1:62,500. Contour interval 10 ft.*

(2) Tidal marsh and tidal streams extend inland into the branching stream valleys between Parker Neck and Bailey Neck and between Custis Neck and Joynes Neck. How can these features be explained in terms of glacial/interglacial changes of sea level. Reconstruct the events of valley erosion and valley filling.

_____

_____

_____

_____

_____

_____

(3) At Metomkin inlet, there is an offset in the straight outer shorelines of Cedar Island and Metomkin Island, the former being farther seaward. Can you explain this offset in terms of shore drifting processes?

_____

_____

_____

_____

_____

(4) Notice that the width of tidal zone (distance from 10-ft contour to the barrier island) decreases from south to north. Give two possible explanations for this widening.

_____

_____

_____

_____

_____

Barrier Beaches of Long Island, New York The south shore of Long Island is yet another fine example of a barrier-island coast. It excels as an example of the shaping of tidal inlets and their relationship to the direction of prevailing shore drift of sand. Figure B shows the principle. Shore drift from east to west allows the updrift end of the barrier to grow rapidly, overlapping the downdrift end, which is narrowed and eroded. The inlet thus migrates downdrift.

Figure C is a simplified map of the western part of Long Island. Notice that Fire Island overlaps Jones Beach, and that Rockaway Beach overlaps Coney Island. However, Jones Beach seems to align quite well with Long Beach across Jones Inlet.

(5) Can you offer an explanation of the lack of overlapping of the barrier beach at Jones Inlet?

_____

_____

_____

_____

_____

_____

_____

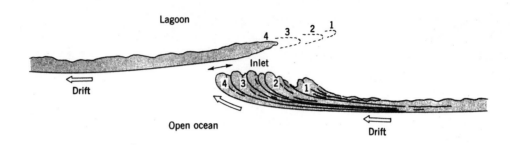

Figure B *Schematic map of migration of an inlet, with offsetting and overlap. (A.N. Strahler.)*

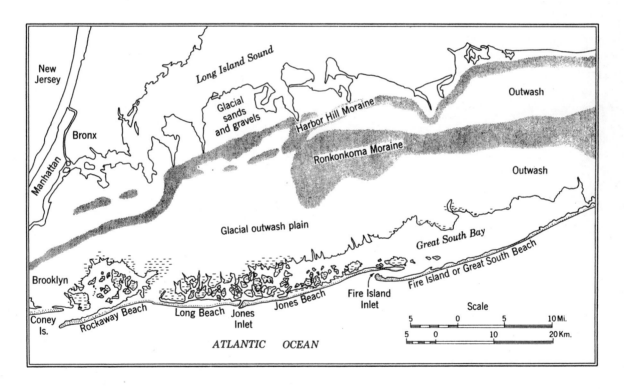

Figure C *Coastal features and moraines of western Long Island, New York. (Based on maps of the U.S. Geological Survey.)*

NAME _____   DATE _____

Exercise 18-D The Ocean Tide

[Text p. 571, Figure 18.9, 18.10.]

For hundreds of millions of years, the ocean tide has played a major role in shaping coastal landforms. In our modern era of high technology, the ocean tide sometimes plays a new and different role. If it should just happen that a great oil tanker with full load, on its way out of the Port of Valdez, Alaska, were to run aground on a submerged rock just after high water of a spring tide, there might be a delay of full lunar cycle of 29 days before the vessel could be floated off that rock. During that period, leaking oil could spread through the branching estuaries of the harbor, carried alternately landward and seaward until the oil reached the shores of the entire estuary system of Prince William Sound. Let's hope that it never happens—but of course, it did happen in 1989.

What causes the tidal rise and fall of the oceans? The moon is in control of the earthly tidal cycle. Figure A shows an imaginary uniform ocean layer over a perfectly spherical solid earth. The force of the moon's gravitational attraction is greatest at point A, closest to the moon; least at C, farthest from the moon; intermediate in strength at B. This diminishing force from A to C causes the tide-raising force, shown by small arrows in Figure B. The ocean water is drawn toward two tidal centers, A and C, where the water tends to accumulate and its level rises to a summit. Thus we have two centers of high water level (at A and C). At the same time, the water subsides to lower level along a great circle passing through the poles. This global "girdle" of low water separates the two centers of high water.

Next, we bring into play the fact that the earth is continually rotating on its axis, so the two bulges and the low girdle continually sweep around the solid earth. An individual at a particular fixed location in low latitudes will observe a rise of ocean level twice each day, alternating with a decline to low level. These are the "high waters" and "low waters," respectively.

The Semi-Daily Tidal Cycle The tidal rhythm explained above gives the kind of tide curve shown in text Figure 18.9. It is called the *semi-daily tidal cycle*. Now, the lunar day—one earth turn with respect to the moon—is about 25 hours (25^h) of mean solar time (clock time). We can anticipate or predict, then, that two successive high waters will occur $12\frac{1}{2}^h$ apart; and same for successive low waters. Does observation confirm this prediction? To find out, we need to plot the data of an actual tidal cycle.

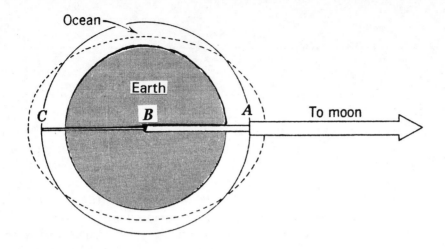

Figure A *Gravitation as the tide-producing force. (A.N. Strahler.)*

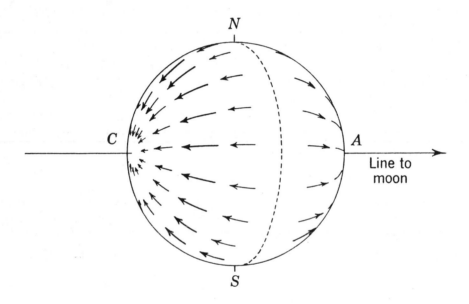

Figure B *Ocean waters tend to move toward two centers, A and C. (A.N. Strahler.)*

Table A San Francisco, California

Hour	Height (Feet)	Hour	Height (Feet)
12 midnight	0.0	1 P.M.	−0.3
1 A.M.	−1.5	2 P.M.	−1.5
2 A.M.	−2.3	3 P.M.	−2.4
3 A.M.	−2.6	4 P.M.	−2.5
4 A.M.	−2.3	5 P.M.	−1.9
5 A.M.	−1.5	6 P.M.	−1.0
6 A.M.	−0.3	7 P.M.	0.1
7 A.M.	0.8	8 P.M.	1.2
8 A.M.	2.0	9 P.M.	2.0
9 A.M.	2.6	10 P.M.	2.3
10 A.M.	2.8	11 P.M.	2.2
11 A.M.	2.3	12 midnight	1.2
12 noon	1.2		

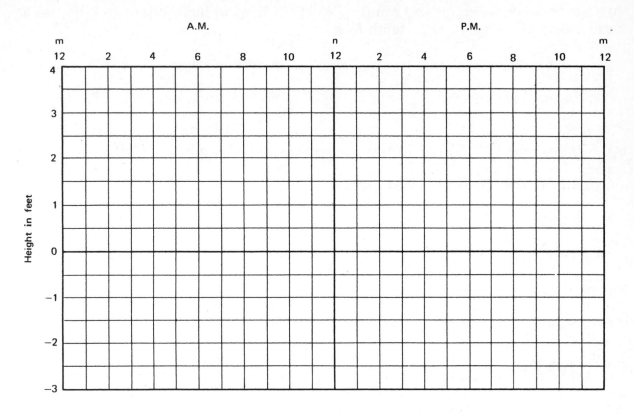

Graph A

(1) Use the data of Table A, showing hourly observations of staff gauge height for San Francisco harbor during a 24-hour period. Plot the data as points on the blank graph (Graph A). Connect the points with a smooth curve.

(2) Our data cover most of two semi-daily tidal cycles, each of about $12\frac{1}{2}^{h}$ duration. Check your plotted tide curve to see if each cycle lasts for $12\frac{1}{2}^{h}$. Actually, the true value of the average semi-daily cycle is more accurately given as $12^{h}\,25^{m}$ (12.42^{h}). Measure the elapsed time between successive low waters and between successive high waters. Calculate the average of the two values.

    Between successive low waters: _____ hrs _____ min

    Between successive high waters: _____ hrs _____ min

            Average: _____ hrs _____ min

Comment on your data. Do they agree with the more exact value given above?

_____

_____

_____

(3) Determine as closely as you can the heights of the two high waters and the two low waters. Give to the nearest one-tenth foot.

First high water _____ ft   First low water _____ ft

Second high water _____ ft   Second low water _____ ft

(4) What were the height ranges between successive low and high waters?

Range from first low to first high: _____ ft

Range from second low to second high: _____ ft

Your answers to Question 4 give the *amplitude* of each tidal cycle. The amplitude is not the same for both cycles in a single lunar day of 25^h, and this is typical of the semi-daily tide curve. Only twice during the lunar month of $29\frac{1}{2}$ days is the amplitude of the two cycles closely matched. This is a topic we must leave unexplained in our exercise, but it has a clear explanation in terms of the path of the moon in the sky, whether high or low.

The Daily Tidal Cycle At various places along the world's coastlines we find that the semi-daily tidal cycle is replaced by a cycle twice as long—the *daily tidal cycle*. Figure C compares these cycles for two stations: Portland, Maine, shows a near-perfect semi-daily cycle; Manila, Philippine Islands, a good example of the daily cycle. Sections of the ocean behave somewhat like tuning forks. For different lengths, tuning forks respond in sympathetic vibration to different sound frequencies. Some sections of the ocean just won't "vibrate" on the two-cycle input, but they vibrate nicely on the daily rhythm. We have just such an example in the tidal cycle for St. Michael, Alaska, a port on the coast of the Bering Sea at about latitude 64°N, just south of the Seward Peninsula.

(5) Plot the data of Table B on Graph B. Connect the points with a smooth curve. Notice that this graph spans two days, or 48 hours, which is twice the time covered in Graph A. The table data are for 2-hour intervals and each graph unit spans 2 hours.

(6) What interval of time elapsed between successive high waters? between successive low waters? (Read to nearest 10 minutes.)

Time of first high water: _____

Time of second high water: _____

Elapsed time: _____

Time of first low water: _____

Time of second low water: _____

Elapsed time: _____

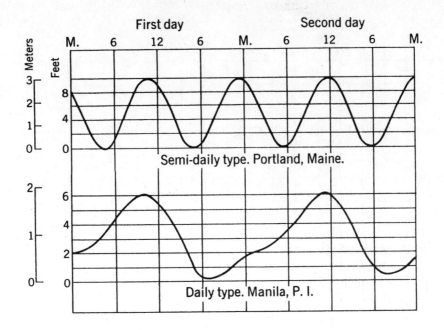

Figure C *Examples of two kinds of tide curves. (A.N. Strahler.)*

Table B *St. Michael, Alaska*

First Day		Second Day	
Hour	Height (Feet)	Hour	Height (Feet)
0	1.0	0	1.1
2	2.0	2	2.3
4	2.3	4	3.0
6	1.7	6	2.9
8	0.4	8	1.9
10	−0.5	10	0.4
12 noon	−1.2	12 noon	−0.5
14	−1.6	14	−1.3
16	−1.8	16	−1.8
18	−1.8	18	−2.2
20	−1.6	20	−2.2
22	−0.5	22	−1.8
24	1.1	24	−0.5

(7) How do your readings compare with a value twice that of the semi-daily cycle, given in Question 2?

_____

_____

_____

_____

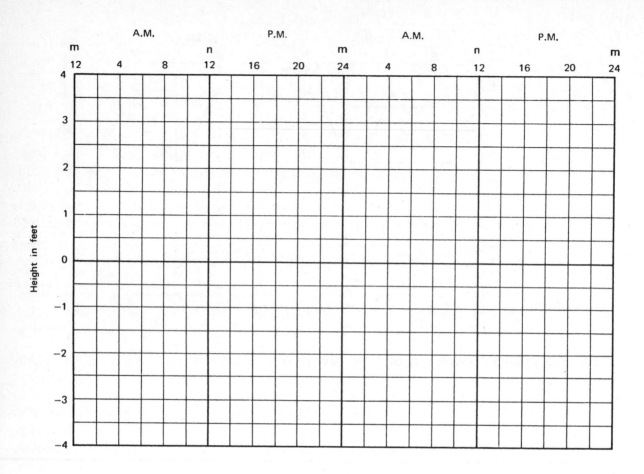

Graph B

(8) Compare the two cycles—semi-daily and daily—as to the uniformity and symmetry of the curves. Look first at Figure A, then at Graphs A and B. Describe your impression of these curves.

_____

_____

_____

_____

_____

NAME _____ DATE _____

Exercise 18-E Cresent Dunes and Sand Seas

[Text p. 579–584, Figures 18.23, 18.24, 18.25.]

For sheer vastness, the North African-Arabian deserts as a group rank first for their huge seas of bare dune sand, over which the sharp dune crests continually shift with the wind. Until satellite imagery became available, North America wasn't world-class in this field, but we now find excellent examples of a wide range of desert dunes in the great Sonoran Desert (Gran Desierto) of northwestern Mexico and the adjacent southwestern United States. Satellite images also reveal new details of vast inland dune fields in the Kalahari Desert of southwestern Africa and deserts in central and western Australia.

In this exercise we investigate those active dunes whose crests lie transverse to the wind—crescent dunes and transverse dunes of sand seas.

Barchans, or Crescent Dunes The isolated barchan dune (or crescent dune) achieves remarkable perfection of symmetry, but can vary considerably in outline and in the proportions of slip face to windward slope. (Pronounce it 'bar-can'.) The variant spelling, *barkhan*, gives a clue to its origin in central Asia. Figure A, showing a side view of barchans along the Columbia River in Oregon, was taken about a century ago by the distinguished American geologist, G.K. Gilbert, and remains unexcelled. Notice that Dr. John Shelton's oblique air photo of a barchan in Peru (text Figure 18.24), shows an almost identical form. That the dune lies upon a rather rough ground surface underlines a salient point about barchans: they must travel downwind. More barchans are shown in Figure B (lower-right corner). Although solitary by nature, barchans also like to link arms as they travel. The little ones travel faster than the large ones, and it is said that the little ones catch up with the big ones, with which they merge.

(1) Using a color pencil or crayon, color several of the barchans in Figure B.

(2) In Figures A and B, and text Figure 18.24, examine closely the desert surface surrounding the barchans. What material composes this floor? Why does it remain unmoved by the same winds that are capable of driving sand over the dune surface? (Clue on text p. 579.)

_____

_____

_____

_____

_____

Figure A *Crescentic (barchan) dunes near Biggs, OR. (G.K. Gilbert, U.S. Geological Survey.)*

Your text describes the manner in which sand grains travel over sand surfaces by leaping and rebounding. The scientific term for this activity is *saltation*, from the Latin *saltare*, to leap or dance. Figure C shows a ground plan and longitudinal profile of a typical barchan. Saltation carries sand grains up the **windward slope** to the **dune crest**, from which the grains fall through the air to land on the **slip face**. Enclosed by the slip face is a sheltered **stagnant air zone** upon which the slip face is encroaching. From each of two **horns**, sand in saltation streams downwind from the dune in a **spray zone**.

(3) On Figure C label all of the terms shown in boldface in the above paragraph. Use arrows as needed to point to the features named. Then complete the circular outline of the barchan to make a complete, symmetrical geometrical figure. What geometrical term describes this outline? Label it on the drawing.

Outline figure: _____

The contour topographic map, Figure D, shows barchans of a large dune field near Moses Lake, Washington.

(4) Using a color pencil or pen, draw in the basal outline of at least twelve well-developed barchans shown on the map. Place a large arrow on the map to show the prevailing wind direction.

(5) Choosing the three highest, best-formed barchans, measure the width, W, (horn-to-horn, across the wind) of each, in feet. Calculate the height, H, of each in feet. Write these figures directly beside the dune. (Note: The grid on the map border is scaled in 1000-yd units.)

Figure B *Sand sea of transverse dunes between Yuma, AZ, and Calexico, CA. (Robert E. Spence.)*

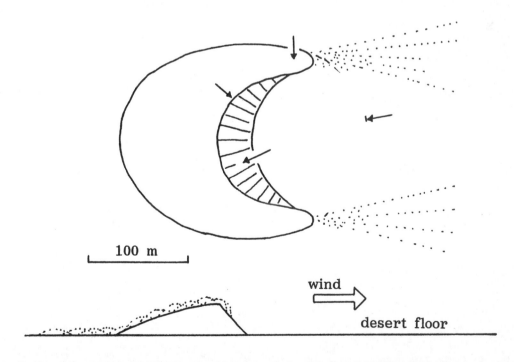

100 m

wind

desert floor

Figure C *Sketch map and profile of an idealized barchan. (A.N. Strahler.)*

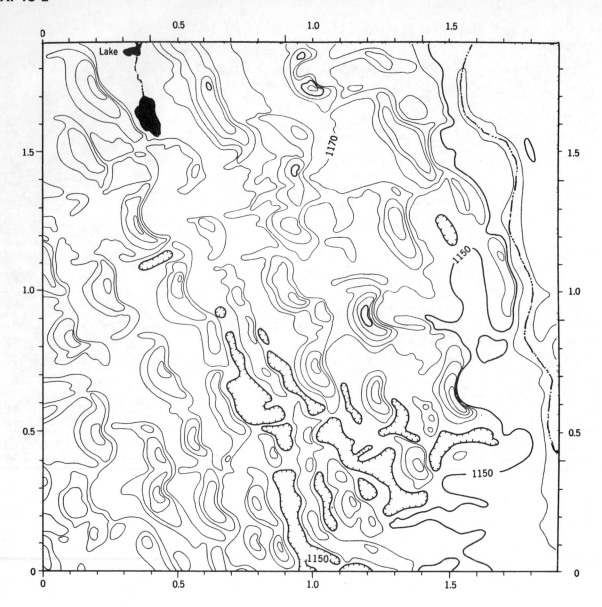

Figure D *Portion of Sieler, WA, Quadrangle. U.S. Geological Survey. Original scale: 1:24,000.*

(6) Barchans close to the western margin of the map show a strange departure from the ideal form shown in Figure C. The horns, instead of pointing straight down-wind, are curved backwards. Find an example at 0.25-1.00. Draw the basal line around this dune to show the special backswept form of the horns. Find and outline in color three other examples of the same dune form. Offer an explanation for this feature. (Hint: text Figure 18.26.)

_____

_____

_____

_____

(7) Study the two small dunes located at 0.60-1.20 and 1.10-1.15. Outline their basal plan. Does their form provide evidence for your explanation given in Question 7.

_____

_____

_____

(8) What evidence is there on the map that the ground-water table lies close to the surface? What influence might this condition have on dune development?

_____

_____

_____

_____

Transverse Dunes of a Sand Sea Figure E is a contour topographic map of the same transverse dunes seen in the oblique air photo, Figure B, and in text Figure 18.25. This field lies between Yuma, Arizona, and Calexico, California. Millions of travelers have seen these dunes from Interstate Highway 8, paralleling the All American Canal (seen in the distance in Figure B) that brings water from the Colorado River to California. Locally, this dune field goes by the name of the Sand Hills.

(9) Compare the two air photos. By looking very closely, you can find the identical dune features on each. How do these photos differ in direction of view and the dune features they emphasize. (Get your directions from the map, Figure E.)

_____

_____

_____

_____

_____

(10) Using color pencil or crayon, shade the slip faces of the dunes. Show direction of prevailing wind with a bold arrow.

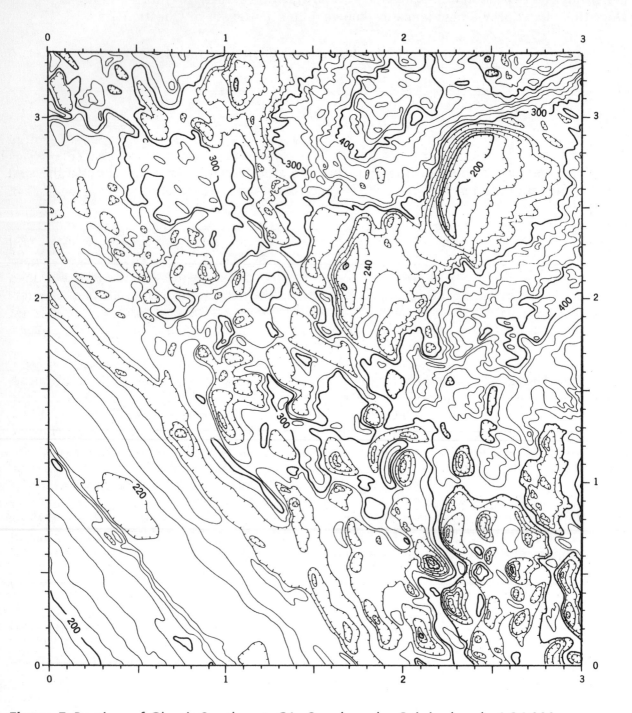

Figure E *Portion of Glamis Southeast, CA, Quadrangle. Original scale 1:24,000.*

Exercise 19-A Living Glaciers of Alaska and British Columbia

[Text p. 594, Figures 19.1, 19.3, 19.5.]

North America has its share of large alpine glaciers. They flourish in a great arc around the Gulf of Alaska: in northern British Columbia, the southwest corner of the Yukon Territory, and across southern Alaska. Here, the combination of lofty mountains ranges and a strong onshore flow of moist marine air guarantees the two essential requirements of alpine glaciers: high mountains and abundant precipitation.

Figure A is a contour topographic map of glaciers in the Chugash Mountains, a coastal range at the head of the Gulf of Alaska. Port cities of Valdez and Cordova are located near here. Valdez, at the head of Prince William Sound, is the terminal point of the Alaska Pipeline. Cordova lies southwest of the map area. The major topographic feature of reference is the Copper River. Its valley lies just off the map to the east, and we see a small piece of its estuary and floodplain in the upper-right corner of the map.

On the original U.S. Geological Survey map, contours on glacial ice are printed in blue, making an easy contrast with the brown contours of the land. A dashed line separates the two kinds of areas.

To give you a better understanding of the glaciated terrain of our map, we include a remarkable air photograph of a similar landscape (Figure B). It shows the great Eagle Glacier and its source icefield in the Coast Mountains of northern British Columbia, near the International Boundary with Alaska. Called a *trimetrogon* photograph, it looks both down and sidewise. When viewing it, imagine that you are in a window seat in an airplane. You can look almost straight down, and also into the far distance. The strange, almost grotesque object you see in the lower right corner of the photo is a huge cirque, containing a small relict glacier. The photo gives a beautiful display of medial moraines. Smooth white glacier surfaces in the distance are firn fields in the zone of accumulation.

(1) Using a color pen or pencil (blue), trace on the topographic map the dashed line that outlines Heney Glacier and the unnamed glaciers that lie east of it on the map. You may wish to complete the tracing over the entire map, or at least over that part including the main tributary glacier system to Heney Glacier.

(2) Using a different color (green or red), trace the drainage divide that surrounds Heney Glacier and the two glaciers to the east of it. This requires careful concentration. A pencil line, lightly drawn, is recommended first, so as to permit easy correction.

(3) Three exceptionally fine large cirques lie at the upper end of Heney Glacier, between 2.0-1.5 and 4.0-4.0. Label these cirques with the letter c. Label three other cirques that lie alone a line from 3.0-6.0 and 6.0-6.0. These belong to north-flowing glaciers tributary to the western branch of Heney Glacier.

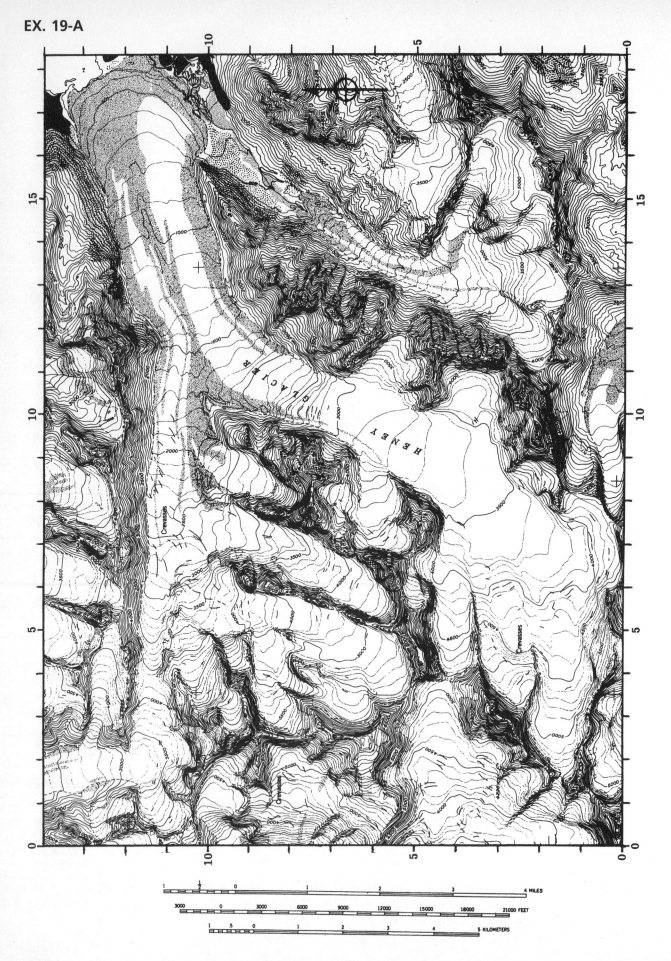

Figure B *Eagle Glacier, British Columbia. (USAF Trimetrogon photo. Canada Department of Mines, No. 99471.)*

(4) Along the divide between the two sets of cirques of Question 3, find and label good examples of a *horn* (**H**), an *arête* (**A**), and a *col* (**CL**). (An arête is the sharp, knifelike divide between two cirques; a col is a low point on an arête).

← **Figure A** *(LEFT) Portion of the Cordova (D-3), AK, Quadrangle. U.S. Geological Survey. Original scale: 1:63,360.*

(5) Study the divide between 11-2 and 17-3 that separates the north-flowing glacier from cirques that face south. Why are the north-facing cirques full of glacial firn and ice, whereas the south-facing cirques are empty?

_____

_____

_____

_____

(6) Follow Heney Glacier downvalley from its cirques to its terminus, noting the spacing of the contours that cross the glacier. Look for changes in gradient. What is the significance of closely crowded contours and associated crevasses (short lines)? Give specific locations by grid coordinates. Explain. (Refer to text Figure 19.3.)

_____

_____

_____

(7) Medial moraines are illustrated and labeled in text Figure 19.5(b). Several are beautifully shown in the text photo, Figure 19.1. Find medial moraines on the map near 9.0-6.0 and 13.0-5.0. Label these **MM** and color them to show their full extent and source points.

(8) Lateral moraines are illustrated and labeled in text figures 19.3 and 19.5(b). Find two good examples on the map. Give grid coordinates and label them **LM**. Color them, same as for the medial moraines.

Grid coordinates: _____ and _____

(9) An _end moraine_ is illustrated and labeled "terminal moraine" in text Figure 19.3. Find two good examples on the map. Give grid coordinates, label them **EM**, and color them, same as for the other kinds of moraines.

Grid coordinates: _____ and _____

(10) Find a good example of a small glacier, occupying a cirque, but not connected to a branch or trunk of a glacier system. Give grid coordinates. Label it **HG**, for "hanging glacier."

Grid coordinates: _____

Exercise 19-B Glacial Landforms of the Rocky Mountains

[Text p. 595, Figures 19.5, 19.7.]

The Middle and Southern Rockies provide many fine examples of landforms of alpine glaciation fully exposed by disappearance of the glaciers that carved them. Two examples that we study in this exercise are from this region; one from the Bighorn Mountains of north-central Wyoming, the other from the Uinta Mountains of northern Utah.

Cirques, Tarns, and Troughs of the Bighorn Mountains For this exercise we use another of those remarkably precise block diagrams, drawn by Erwin Raisz (Figure A), to give a vivid picture of the terrain shown on the contour topographic map, Figure B.

(1) Determine the contour interval of the map. Measure precisely the width of the area it covers.

Contour interval: ___200___ ft.

Width of map area: ___11,000___ yds; ___6.25___ mi.

(2) This is a good contour map on which to practice the art of plastic shading to bring out the visual effect of surface relief. Using a very soft lead pencil, apply shading to steep slopes that face south, southeast, and east. Make the shading darkest where the slope is steepest, as in the headwalls of the cirques. Make a sharp edge on the dark shading to reveal the sharp break between the gently sloping mountain summit upland and the cirque headwall.

Each small lake in a cirque and its trough occupies a rock basin, scoured out by the glacier, although in some cases a moraine may form a dam that holds a lake. Such rock-basin lakes are sometimes called *tarns*, an ancient Scandinavian term. Figure C shows a splendid rock-basin lake in the Northern Rockies of Glacier National Park. It occupies a cirque with steep walls rising more than 4,000 ft to the surrounding horns and aretes. Chains of tarns, such as those on your map, are sometimes called "paternoster lakes" from their resemblance to beads of a rosary.

Troughs of the Bighorn Mountains look very much like the one shown in text Figure 19.7, which lies in the nearby Beartooth Mountains, a geologically similar uplift with a crystalline rock core into which the glacial features are carved. Notice that the upland on either side of the trough is a rolling, plateau-like surface. Both regions are described as being in a "youthful" stage of alpine glaciation, as shown in the right-rear corner of text Figure 19.5(b).

Figure A *Block diagram of landforms of the Bighorn Mountains included the topographic map, Figure B. (Drawn by Erwin Raisz.)*

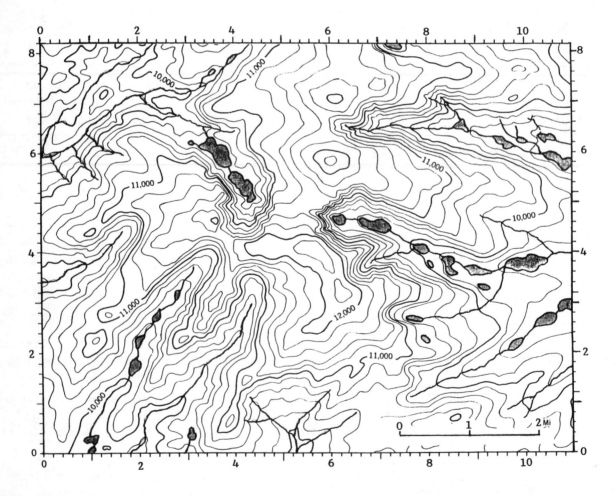

Figure B *Portion of the Cloud Peak, WY, Quadrangle. U.S. Geological Survey. Original scale: 1:125,000.*

Figure C *Lake Ellen Wilson in Glacier National Park, Montana. (Douglas Johnson.)*

(3) Using a color pencil or pen, trace on the map (Figure B) the main drainage divide of the mountain range. Label it **divide**. Explain why cirques heading on the east side of the range have eaten back closer to the main divide than those heading on the west side of the range. You may need to speculate as to the cause. (Hint: Think about (a) winds and (b) insolation.)

_____

_____

_____

_____

_____

(4) Using a color pencil, lay out on the map the shortest possible route for a trail across the mountain range from 5.5-2.3 to 0.0-1.5 in such a way that the gradient nowhere exceeds 1,600 ft per mi. Avoid all cirques and troughs.

Cirques, Arêtes, and Horns of the Uinta Mountains The example of glaciated mountains in the Unita range, shown in Figures D, E, and F is strikingly different from that in the Bighorn Range. Like the Bighorn range, the Uinta range is a great arch, or anticline, in which sedimentary strata were uplifted in the Laramide Revolution that closed the Mesozoic Era. The top of the arch is flattened. Whereas the Bighorn arch has been eroded to the point that a large crystalline rock core is exposed, the Uinta arch still retains strata that pass over the summit. These strata are quartzite formations of Precambrian age. Weaker formations within the quartzite layers allowed high, steep cliffs of quartzite to retreat as the cirques enlarged, much as the Redwall and Coconino cliffs have retreated in the walls of Grand Canyon. Figure F is a very old photograph showing the high, steep quartzite cliffs surrounding a cirque with its rock-basin lake. Notice the succession of cliffs and ledges. (This cirque is just off the top of your map.)

(5) Write the word cirque on several good examples on the map. Label examples of arêtes, horns, and cols.

(6) Color blue all areas on the map that you would suppose were covered by glacial ice at a stage of maximum glaciation. Add numerous arrows to show directions of ice movement. In another color, draw in all medial moraines that you might expect to have existed on the glaciers.

(7) Summarize the major differences between the Uinta glacial landforms and those of the Bighorn range.

_____

_____

_____

_____

_____

_____

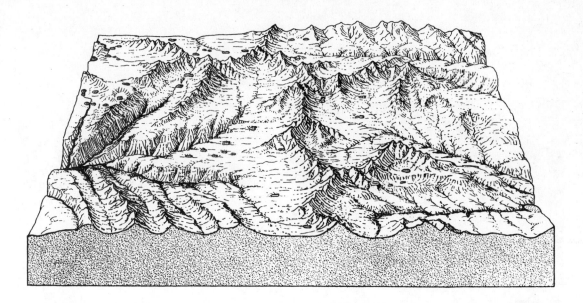

Figure D *Block diagram of landforms of the Uinta Mountains included in the topographic map, Figure B. (Drawn by Erwin Raisz.)*

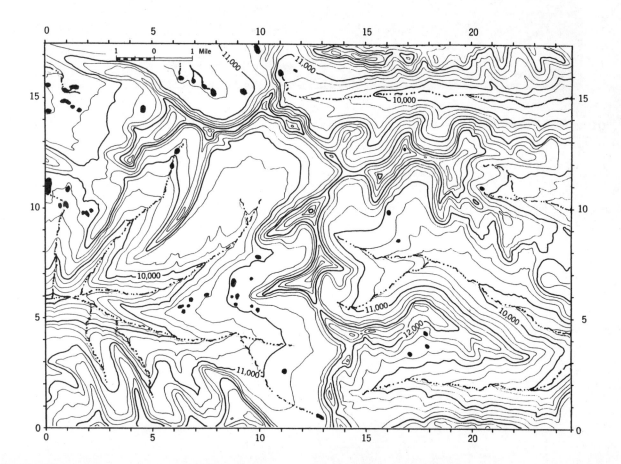

Figure E *Portion of the Hayden Peak, UT, Quadrangle. U.S. Geological Survey. Original scale 1:125,000. North is to the right.*

347

Figure F *A cirque in the Uinta Mountains. (Wallace W. Atwood, U.S. Geological Survey, 1909.)*

NAME _____ DATE _____

Exercise 19-C Moraines and Outwash Plains

[Text p. 603–606, Figure 19.18.]

Over much of the northcentral and northeastern United States, and extending far north into Canada, landforms shaped by the great Wisconsinan ice sheets and the meltwaters streaming from them are today scarcely touched by weathering and mass washing and the eroding forces of runoff. But humans have heavily impacted these land forms in and around our urban areas—excavating them and regrading them for highways and shopping malls. Try to find a good exposure of a glacial delta to visit on a field trip, and you will soon get the point. The beautifully washed and graded sand of the delta is now elsewhere, bound in concrete. Outwash plains, also deposits of sand and gravel, are mostly concealed under houses, lawns, and pavements of a thousand housing developments.

The total complex of landforms found near what was formerly the terminal or recessional belt of an ice sheet is laid out in the two block diagrams of text Figure 19.18. To understand how these landforms take expression in topographic contours, we use a map of an area along the Wisconsin-Minnesota boundary.

(1) Study the map, Figure A, alongside diagram *b* of text Figure 19.18. Look closely at the lower left end of the block and the cross section it bears. Match the various kinds of landforms with their locations on the map. Label examples each of the following on the map:

Moraine	Kettle lake
Knob-and-kettle	Ground moraine
Outwash plain	Drumlin
Ice-block depression	Basal glacial till

(2) Compare the surface elevation of the outwash plain close to the moraine with that of the surface just north of the moraine. Explain the relationship in elevations of these two surfaces.

_____

_____

_____

_____

_____

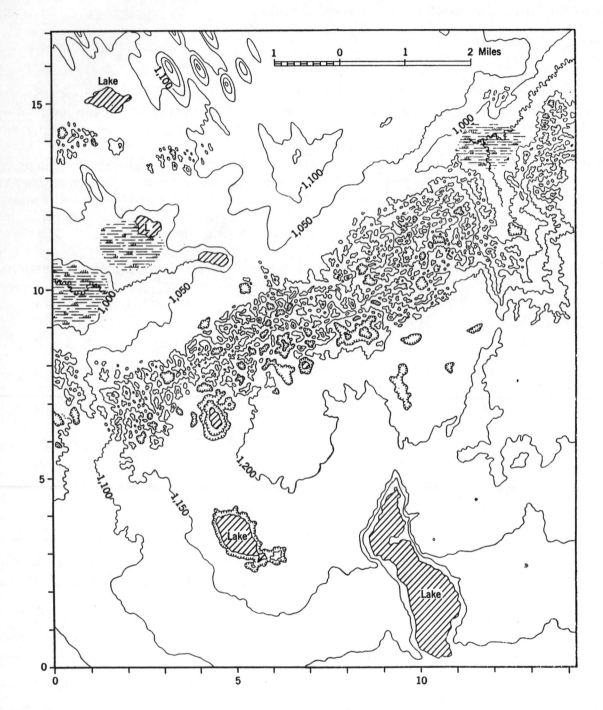

Figure A *Portion of the St. Croix Dalles, WI-MN, Quadrangle. U.S. Geological Survey. Original scale 1:62,500.*

(3) How can you explain the presence of ice-block depressions (kettles) in the outwash plain lying well south of the stagnant ice margin, where the moraine was being deposited?

_____

_____

_____

_____

(4) Draw a bold arrow on the northern part of the map to show the direction of ice movement. What two kinds of evidence suggest the flow direction? Are they in agreement?

_____

_____

_____

_____

(5) What is the origin of the basins occupied by lakes and marshes north of the moraine?

_____

_____

_____

_____

(6) Study the group of small hills and depressions lying between 0-14 and 4-14. How do you interpret them in terms of features of glaciation?

_____

_____

_____

_____

(7) If you planned to excavate sand and gravel for commercial use in the production of concrete, where on this map would you locate the pit? Indicate by one or more squares labeled **sand/gravel pit**. Explain your choice.

_____

_____

_____

_____

NAME _____ DATE _____

Exercise 19-D Eskers and Drumlins

[Text p. 605, Figures 19.18, 19.20, 19.21.]

This exercise brings together two very odd landforms of unlike origins—eskers and drumlins—although both originate beneath an ice sheet. Eskers are stream deposits in subglacial tunnels in stagnant ice; drumlins are till deposits plastered over the subglacial floor by moving ice. It stands to reason that eskers and drumlins cannot be formed in the same place at the same time, but it would be possible for an esker to run right over a drumlin, if the esker formed after the moving ice had become completely stagnant.

Eskers The word *esker* is said to be of Gaelic-Irish origin, and you will find it in their literature spelled also as *eskir, eskar, excar*, and *eiscir*. Sweden has many fine eskers, which the Swedes call "osars." A more descriptive name used in New England is "horseback," and indeed, the crest of an esker would provide a fine horse trail well above adjacent bug-ridden swamps and muskeg—all the better because the crest of an esker is typically free of forest trees. Canada must have more eskers than any other nation; they are found from Labrador on the east to Great Bear Lake on the west, forming fanlike patterns that show clearly the flow directions within the great ice sheets.

Figure A is an oblique air photo of an esker in the Canadian shield It is an old photo, but they don't come much better. It shows the heavily glaciated Precambrian crystalline rocks with their numerous ice-scoured basins. Notice that the esker crosses the lake basins as well as the hill summits between them, almost as if dropped at random across the terrain. Actually, eskers form on or close to the rock floor and the water in the ice tunnel can flow uphill as well as downhill, depositing gravel as it goes. Eskers show thinning and thickening of their height and width, looking a bit like a python after a big meal. The esker crest in this photo appears highly reflective of light, because of the scarcity of trees on the crest.

Map A in Figure B is a contour topographic map of an esker in Maine, known locally as the Enfield Horseback. The contour interval is 20 ft.

(1) Estimate the width of the esker just north of the Passadumkeag River. How high is it above the surrounding plain?

Width: _____ to _____ yds. Height: _____ to _____ ft.

Figure A *An esker near Boyd Lake, Canada. (Canada Dept. of Mines, Geological Survey, Atlas 126.)*

(2) Why does the esker crest rise and fall in elevation along its length?

_____

_____

_____

(3) Explain the rise in elevation of the esker crest in the vicinity of 3.3-6.0.

_____

_____

_____

(4) Explain the very small closed depressions at 2.3-13.9.

_____

_____

_____

Drumlins The term *drumlin* is from Ireland, where it is used to describe various kinds of rounded hills, and has been adopted internationally as a scientific term for a subglacial deposit of till shaped by the ice flow into a streamlined form. County Down in northeast Ireland contains nearly 4,000 bonafide drumlins. The drumlin in text Figure 19.21 shows what is often considered the ideal drumlin form. (Read the figure caption.) New York claims the largest American drumlin field, with some 10,000 being counted along the south shore of Lake Ontario. To the north of that lake, the province of Ontario has four important drumlin fields. Drumlins of a great field in east-central Wisconsin number about 5,000, and there are many in New England and Nova Scotia, as well.

From the viewpoint of American history, however, Boston's drumlins have the highest distinction. Two that lie close to the historic heart of the city are Breeds Hill and Bunker Hill. In May of 1775, colonial militia had laid siege to the British forces in Boston. General Howe then arrived with reinforcements with the intention of breaking the siege, and with rumored plans of taking the heights of Charlestown. To counter this threat, the Continental commander, William Prescott, was sent to fortify Bunker Hill. Instead, he chose the other drumlin, Breeds Hill, as his position and there in June fought a furious and bloody battle before being forced to withdraw. But, as history would have it, this was the Battle of Bunker Hill.

Boston Harbor has many fine drumlins, several of which are shown in Map B of Figure B. They are more rotund in outline than the classic examples from farther inland, and you will need to explain that feature.

(5) Measure the length, width, and height of the drumlin at 1.2-2.7.

Length: _____ ft.   Width: _____ ft.   Height: _____ ft.

(6) Using the long axes of the drumlins as an directional indicator of ice movements, compare the directions indicated by the drumlins in the larger map (Nantasket Beach, Telegraph Hill, etc.) and those on Paddocks Island. Paddocks Island actually lies only a short distance due west of Telegraph Hill. Offer an explanation of the differences in direction you observe.

_____

_____

_____

_____

_____

(7) Drumlins of western New York State are shown in Map C of Figure B. How many drumlins are shown on this map?

Number of drumlins: _____

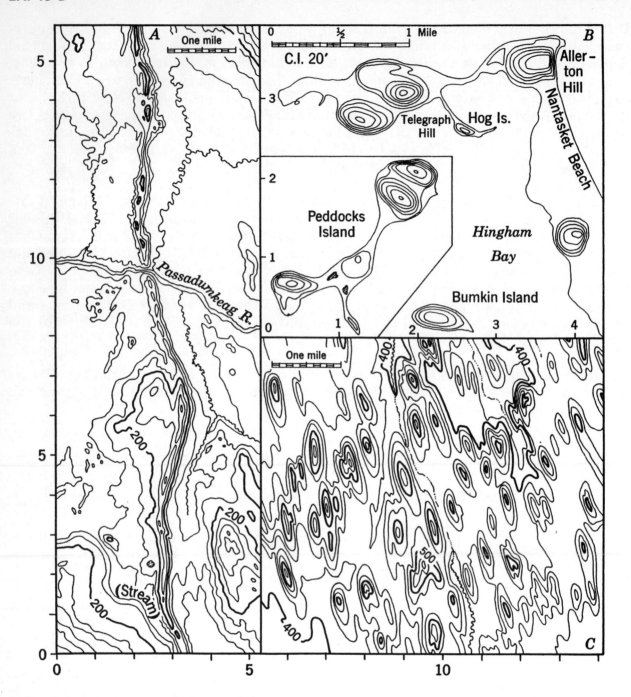

Figure B *Portions of topographic maps of the U.S. Geological Survey.*
(A) Passadumkeag, ME, Quadrangle. Original scale 1:62,500. (B) Hull, MA, Quadrangle.
Original scale 1:31,680. (C) Weedsport, NY, Quadrangle. Original scale 1:62,500.

(8) Estimate the direction (azimuth) of ice flow indicated by this group of drumlins.

Azimuth: _____

Figure C *Vertical air photograph of drumlins and till grooves near Carp Lake, northern British Columbia, (U.S. Air Force.)*

(9) The drumlins of Map B are somewhat differently shaped than those on Map C. Describe and explain this difference.

_____

_____

_____

_____

Figure C is a vertical air photograph of drumlins near Carp Lake in northern British Columbia. The area shown is about 4 mi wide. The drumlin forms are side by side with long narrow grooves and ridges, also formed in glacial till. Some of the drumlins show narrow grooves on their surfaces. In some areas of Canada the land surface consists entirely of straight, parallel grooves with intervening narrow ridges, and is called a "fluted till surface."

(10) Using a color pen, outline several well-shaped drumlins. Draw lines to emphasize a set of well-developed parallel grooves.

(11) Using a ruler or triangle, draw a slanting line across the photograph to indicate the average direction of motion of the ice when it formed these drumlins. Assuming that the photograph is correctly oriented with respect to geographic north, measure the direction of ice motion. Give your reading as both a compass quadrant bearing and as azimuth. (See Appendix, Figure A.13.)

Bearing: _____   Azimuth: _____

APPENDIX

Topographic Map Reading

Several methods are used to show accurately the configuration of the land surface on topographic maps. These are plastic shading, altitude tints, hachures, and contours (Figure A.1). The first three methods give a strong visual effect of three dimensions so that most persons can easily grasp the essential characteristics of the landscape features. These methods of showing relief are, however, inadequate because they do not tell the reader the elevation above sea level of all points on the map or how steep the slopes actually are. Topographic contours give this information and make the most useful type of topographic map.

An example of a modern contour topographic map will be found in the back end pages of your textbook, *Introducing Physical Geography*. This example represents the prevailing style of large-scale topographic maps issued by the U.S. Geological Survey.

Plastic Shading

Maps using *plastic shading* to show relief look very much like photographs taken looking down on a plastic relief model of the land surface (Figure A.1). The effect of relief is produced by gray or brown tones applied according to the oblique illumination method. Light rays are imagined as coming from a point in the northwestern sky somewhere intermediate between the horizon and zenith. Thus all slopes facing southeast receive the heaviest shades and are darkest where the slopes are steepest. Maps with plastic shading are much like air photographs, looking straight down (Figure A.2).

Altitude Tints

In its simplest form, *altitude tinting* consists of assigning a certain color, or a certain depth of tone of a color, to all areas on the map lying within a specified range of elevation. Maps in atlases and wall maps commonly show low areas in green, intermediate ranges of elevations in successive shades of buff and light brown, and high mountain elevations in darker shades of brown, red, or violet. Each shade of color is assigned a precise elevation range. The method is effective for small-scale maps that are viewed at some distance.

Hachures

Hachures are very fine, short lines arranged side by side into roughly parallel rows. Each hachure line lies along the direction of the steepest slope

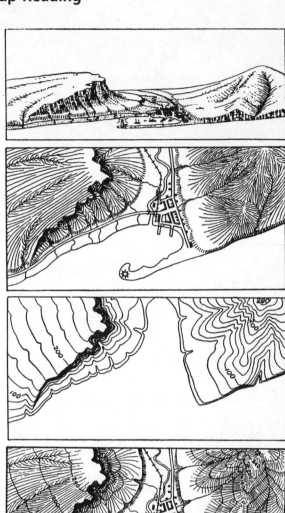

Figure A.1 Various ways in which relief can be shown are, from top to bottom: (1) perspective diagram or terrain sketch, (2) hachures, (3) contours, (4) hachures and contours combined, and (5) plastic shading and contours combined (Drawn by E. Raisz.)

359

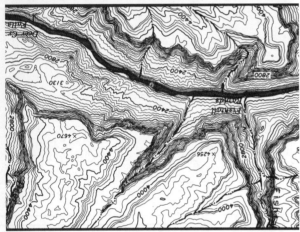

Figure A.2 A vertical air photograph (above) serves as a kind of topographic map, showing all relief details but lacking elevation information. The contour topographic map (below) covers the same area. It shows the Colorado River flowing in the bottom of the Grand Canyon. Short tributary streams have carved side canyons in massive cliff-forming strata. North is toward the bottom of both photograph and map to give the proper effect of relief in viewing the photograph. (U.S. Forest Service and U.S. Geological Survey.)

Contour Lines

A *contour line* is an imaginary line on the ground, every point of which is at the same elevation above sea level. Contour lines on a map are simply the graphic representations of ground contours, drawn for each of a series of specified elevation intervals, such as 10, 20, 30, 40, or 50 meters (or feet) above sea level or above any other chosen base, known as a *datum plane*. The resulting line pattern not only gives a visual impression of topography to the experienced map reader, but also supplies accurate information about elevations and slopes.

Figure A.4. The shoreline is a natural contour line because it is a line connecting all points having zero elevation. Suppose the sea level could be made to rise exactly 10 meters (or that the island could be made to sink exactly 10 meters); the wa-

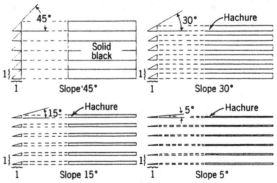

Figure A.3 A portion of the Metz sheet is shown above to correct scale. This map is one of the French 1:80,000 topographic series using black hachures and spot heights. The Lehmann system of hachuring, shown below, varies the thickness of hachure line according to ground slope.

and represents the direction that would be taken by water flowing down the surface.

A precise system of hachures, adapted to representation of detailed topographic features on accurate, large-scale maps, was invented by J.G. Lehmann and was widely used on military maps of European countries throughout the nineteenth century. In the Lehmann system, steepness of slope is indicated by thickness of the hachure line (Figure A.3).

Because hachures do not tell the elevation of surface points, it is necessary to print numerous elevation figures on hilltops, road intersections, towns, and other strategic locations. These numbers are known as *spot heights*. Without them a hachure map would be of little practical value.

360   Copyright © by Arthur N. Strahler

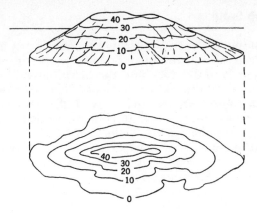

Figure A.4 Contours on a small island. (Drawn by A.N. Strahler.)

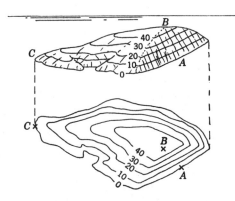

Figure A.5 On the steep side of this island, the contours appear more closely spaced. (Drawn by A.N. Strahler.)

short horizontal distance and therefore appear closely spaced on the map. To travel from B to the shore at C requires the same total vertical descent; but, because the slope is gentle, the horizontal distance traveled is much greater and the contours between B and C are widely spaced on the map.

Selection of a contour interval depends both on relief of the land and on scale of the map. Topographic maps showing regions of strong relief require a large interval, such as 10, 25, or 50 m (50, 100, or 200 ft); regions of low to moderate relief require such intervals as 2 or 5 m (10 or 25 ft).

Because much of the earth's land surface is sculptured by streams flowing in valleys, special note should be made of how contours change direction when crossing a stream valley. Figure A.6 is a small contour sketch map illustrating some stream valleys. Notice that each contour is bent into a "V," the apex of which lies on the stream and points in an upstream direction.

Determining Elevations by Means of Contours

Figure A.6 can be used to illustrate the determination of elevations. Point B is easy to determine because it lies exactly on the 1300 m contour. Point C requires interpolation. Because it lies midway between the 1100 and 1200 m lines, its elevation is estimated as the midvalue of the vertical interval, or 1150 m. Point D lies about one-fifth of the distance from the 1000 m to the 1100 m contour. One-fifth of the contour interval is 20 m. Thus we estimate that point D has an approximate elevation of 1020 m. If the ground is not highly irregular, the error of estimate will probably be small. Determination of the summit

ter would come to rest along the line labeled "10." This new water line would be the 10-meter contour because it connects all points on the island that are exactly 10 meters higher than the original shoreline. By successive rises in water level, each exactly 10 meters more than the last, the positions of the remaining contours would be fixed.

Contour Interval and Slope

Contour interval is the vertical distance separating successive contours. The interval remains constant over the entire map, except in special cases where two or more intervals are used on the map sheet.

Because the vertical contour interval is fixed, horizontal spacing of contours on a map varies with changes in land slope. The rule is that close crowding of contour lines represents a steep slope; wide spacing represents a gentle slope. Figure A.5 shows a small island, one side of which is a steep, clifflike slope. From the summit point B to the cliff base at A, the contours are crossed within a

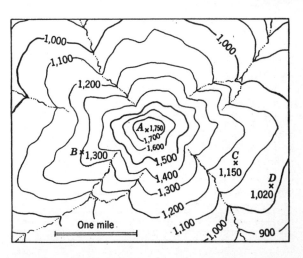

Figure A.6 Stream valleys produce V-shaped indentations of the contours. Elevations are given in meters.

elevation, point A, involves more uncertainty. It is certain that the summit point is more than 1700 m and less than 1800 m because the 1700 m contour is the highest one shown. Because a fairly large area is included within the 1,700 m contour, we may suppose that the summit rises appreciably higher than 1700 m. A guess would place the actual elevation at about 1750 m.

On many topographic maps the elevations of hilltops, road intersections, bridges, streams, and lakes are printed on the map to the nearest meter or foot. These spot heights do away with the need for estimating elevations at key points.

Government agencies, such as the U.S. Geological Survey and the U.S. Coast & Geodetic Survey, determine the precise elevation and position of convenient reference points. These points are known as *bench marks*. On the map they are designated by the letters *B.M.*, together with the elevation stated to the nearest meter or foot.

Depression Contours

A special type of contour is used where the land surface has basinlike hollows, or *closed depressions*, which would make small lakes if they could be filled with water. This contour line is the *hachured contour*, or *depression contour*. Figure A.7 is a sketch of a depression in a gently undulating plain. Below the sketch is the corresponding contour map. Hachured contours have the same elevations and contour intervals as regular contours on the same map.

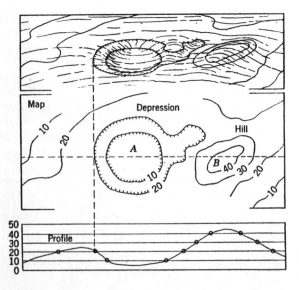

Figure A.7 Contours that close in a circular pattern show either closed depressions or hills.

Map Scale

Distance between points shown on a map depends on the *scale* of the map, defined as the ratio between map distance and the actual ground distance that the map represents. This *fractional scale* (*representative fraction*, or *R.F.*) can be converted to conventional units of length. For example, a map scale of "1:100,000" can be read as "one centimeter represents one kilometer." Thus an ordinary centimeter scale can thus be used to measure map distances.

Most maps of small areas carry a *graphic scale* printed on the map margin. This device is a length of line divided off into numbered segments (Figure A.8). The units are in conventional terms of measurement, such as meters and kilometers or feet, yards, and miles. To use the graphic scale, hold the edge of a piece of paper along the line to be measured on the map and mark the distance on the edge of the paper. Then place the paper along the graphic scale and read the length of the line directly.

Topographic Profiles

To visualize the relief of a land surface, *topographic profiles* can be drawn. These are lines that show the rise and fall of the surface along a selected line crossing the map. Figure A.9 shows how to construct a profile. Draw a line *XY*, across the map at the desired location. Place a piece of paper, ruled with horizontal lines, so that its top edge lies along the line *XY*.

Each horizontal line represents a contour level and is so numbered along the left-hand side. Starting at the left, drop a perpendicular from the point a where the map contour intersects the profile line, *XY*, down to the corresponding horizontal level. Mark the point *a'* on the horizontal line. Next, repeat the procedure for the 400 m contour at point *b*, and so on, until all points have been plotted. Then draw a smooth line through all the points, completing the profile. Where contours are widely spaced, some judgment is required in drawing the profile.

Figure A.9 shows two profiles, both of which are drawn along the same line *XY*. The difference is one of degree of exaggeration of the vertical scale. In this illustration, the horizontal map scale is 1 cm = 1000 m, or 1:100,000, whereas the vertical scale of the upper profile is 1 cm = 100 m, or 1:10,000. The vertical scale is thus 10 times larger than the horizontal map scale, and the profile is said to have a *vertical exaggeration* of 10 times (×10). In the lower profile the horizontal scale remains the same of course, but the vertical scale is 1 cm = 200 m, or 1:20,000. The vertical exaggeration is therefore 5 times (×5). Some degree of vertical exaggeration is usually needed to bring out the details of the topography.

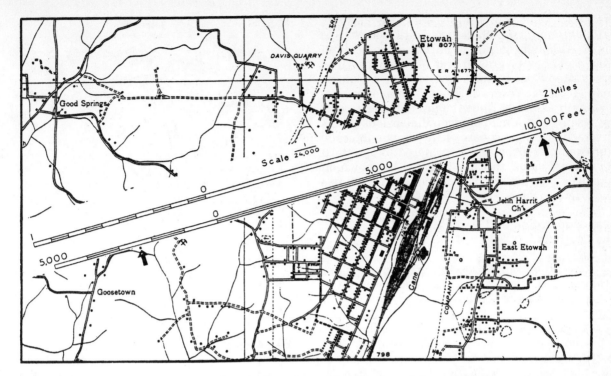

Figure A.8 Portion of a large-scale map for which two graphic scales have been provided. (U.S. Geological Survey.)

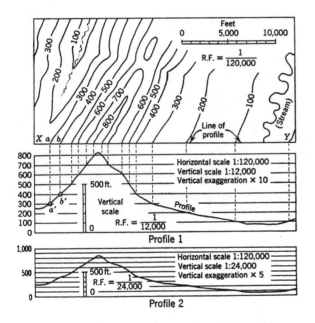

Figure A.9 Construction of a topographic profile.

Large-Scale and Small-Scale Maps

The relative size of two different scales is determined according to which fraction (R.F.) is the larger quantity. For example, a scale of 1:10,000 is twice as large as a scale of 1:20,000. Maps with scales ranging from 1:600,000 down to :100,000,000 or smaller are known as *small-scale maps*; those of

scale 1:600,000 to 1:75,000 are *medium-scale maps*, and those of scale greater than 1:75,000 are *large-scale maps*.

For representing details of the earth's surface, large-scale maps are needed and the area of land surface shown by an individual map sheet must necessarily be small. A topographic sheet measuring 40 cm by 50 cm, on a scale of 1:100,000 (1 cm = 1 km), would of course include an area 40 km by 50 km or 2000 square kilometers (2000 km²). Of the common sets of topographic maps published by national governments for general distribution, most fall within the scale range of 1:20,000 to 1:250,000.

Relationship Between Scales and Areas

Assuming that two maps, each on a different scale, have the same dimensions, what is the relationship between the ground areas shown? Figure A.10 shows three maps, each having the same

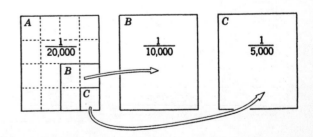

Figure A.10 Map area decreases as scale increases.

363

dimensions but representing scales of 1:20,000, 1:10,000, and 1:5,000, respectively, from left to right. Although map *B* is on twice the scale of Map *A*, it shows a ground area only one-fourth as great. Map *C* is on four times as large a scale as map *A*, yet it covers a ground area only one-sixteenth as much. Thus the ground area that is represented by a map of given dimensions varies inversely with the square of the change in scale. For example, if the scale is reduced to one-third its original value, the area that can be shown on a map of fixed dimensions increases to nine times the original value.

Map Orientation and Declination of the Compass

By convention, large-scale topographic maps are oriented so that north is in a direction toward the top of the map and south is toward the bottom of the map.

The geographic north pole, to which all meridians converge, forms a reference point for *true north* or *geographic north*. There is, however, another point on the earth, the *magnetic north pole*, to which compasses point (Figure A.11). The magnetic north pole is located in the Northwest Territories of Canada, among the Queen Elizabeth Islands, at about lat. 80°N, long. 100°W. It has been moving poleward at an average rate of about

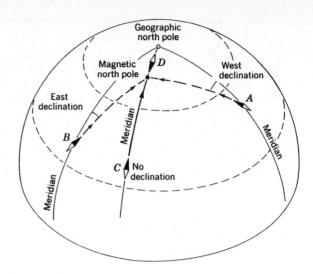

Figure A.11 Whether declination is east or west depends on the observer's global position with respect to the magnetic and geographic north poles.

24 km (15 mi) per year. Most large-scale maps have printed on the margin two arrows stemming from a common point. One arrow designates true north; the other, *magnetic north*. The angular distance between the two directions is known as the *magnetic declination*.

Magnetic declination varies greatly in different parts of the world, depending principally on

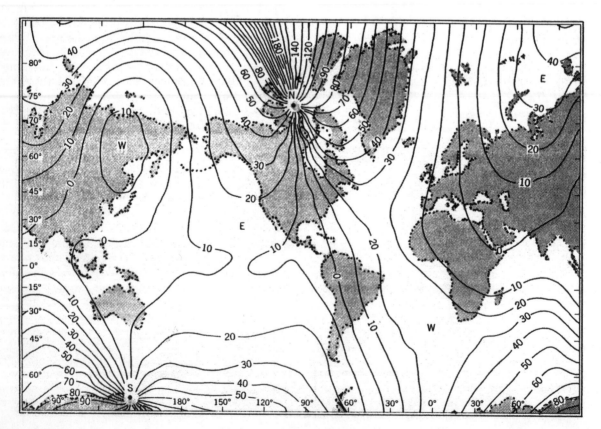

Figure A.12 On this world isogonic map, declination is given in intervals of 10°. (Data of U.S. Naval Oceanographic Office.)

one's position relative to the geographic and magnetic poles. Lines on a map drawn through all places having the same compass declination are known as *isogonic lines* (Figure A.12). The line of zero declination (agonic line) runs through eastern North America. Anywhere along this line, the compass points to true geographic north and no adjustment is required.

Magnetic declination changes appreciably with the passage of years. The amount of annual change in declination is usually stated on the margin of the map.

Bearings and Azimuths

When a map is used in the field, it is often necessary to state the direction followed by a road or stream or to describe the direction that can be taken to locate a particular object with respect to some known reference point. For this purpose, the observer determines the horizontal angle between the line to the objective and a north-south line. The common unit of angular measurement is the *degree*, 360 of which comprise a complete circle. Other systems of angular measurement, such as the *mil* (of which there are 6400 in a complete circle), are sometimes preferable for special applications.

Two systems are used to state direction with respect to north. *Compass quadrant bearings* are angles measured eastward or westward of either north or south, whichever happens to be the closer. Examples are shown in Figure A.13A. The direction from a given point to some object on the map is written as "N49°E" or "S70°W." All bearings range between 0 and 90°. Compass bearings may be *magnetic bearings*, related to magnetic north, or *true bearings*, related to geographic north. Unless otherwise stated, a bearing should be assumed to be a true bearing.

Azimuths are used by military services and in air and marine navigation generally. As shown in Figure A.13B, all azimuths are read in a clockwise direction from north and range between 0° and 360°. Azimuths are measured from either magnetic north or true north, referred to as *magnetic azimuth* and *true azimuth*, respectively.

Map Coordinate Systems

Any system used to locate points on the earth's surface with reference to a fixed set of intersecting lines is a *coordinate system*. The global system of geographic coordinates using parallels of latitude and meridians of longitude is described in your textbook, Chapter 1. Exercise 1-A of this manual deals with that subject.

Another global system is the *Universal Transverse Mercator (UTM) grid system*, used by the United States armed services. It is a rectangular grid superimposed on a special variety of the Mercator projection. Figure A.14 shows how the UTM

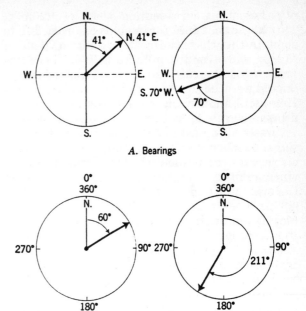

A. Bearings

B. Azimuths

Figure A.13 Directions are expressed as bearings or azimuths.

grid system is set up. It uses the meter as the basic unit of length. The grid is essentially a network of squares, each square 1000 m wide. A portion of a 1000-meter grid is shown in Figure A.14. Numbers on vertical grid lines increase eastward, toward the right; those on horizontal grid lines increase northward, or upward. Only two digits are printed on most grid lines: those that denote thousands and tens of thousands. In the case of the 1000-meter grid, three zeros have been dropped, as well as those digits that denote hundreds of thousands and millions. Full numbers are printed once near the lower lefthand corner of the map. Grid line spacing of 1000 m is used on large-scale maps 1:100,000 or smaller. On the back end paper of your textbook you will find the coordinates of this grid printed along the right-hand and bottom edges of the quadrangle map.

In stating grid coordinates, the number of meters east (right) is given first; then the number of meters north (up). This gives a simple rule: "Read right up." In giving the grid coordinates of a point on the map, the first step is to determine the 1000-meter grid square in which the point lies. A grid square is designated by the grid coordinates of its lower left-hand corner. Thus grid square *A* in Figure A.14 is designated by the intersection of grid lines *87* east and *80* north. These numbers are written together as *8780*. This is a shorthand notation for coordinates 687,000 m east, 3,800,000 m north.

For a particular point within a grid square, the coordinates may be read to the tenth part of a grid square, which is 100 m. Point *B* of Figure

A.14 lies about four-tenths of the distance from 84 to 85, so that one coordinate is *844* east. It lies about five-tenths of the distance up from 76 to 77, so that the second coordinate is *765*. These are written together as a six-digit number, 844765. Should we need to locate the point to the nearest 10 m, still another digit is read by more precise measurements. Thus Point *C* is found to have coordinates *8715* east and *7783* north; written as a single number:87157783. In all three examples, above, an even number of digits forms the combined number. Therefore, given a coordinate designation, the number must be broken into two halves. The first half is taken as the *easting*; the second half as the *northing*.

Researchers in many branches of the geosciences—including physical geography and geology—are now specifying in their research reports the precise locations of objects in the field in UTM grid coordinates. Thus a new level of precision is provided in scientific description of phenomena.

Using the UTM grid system requires a special kind of north: *grid north* (GN). This leads to a definition of a second kind of declination, *grid declination*, which is the angle between *grid north* (the direction taken by the vertical grid lines) and true (geographic) north. Grid declination can be read directly with a protractor placed on the angle formed between a grid line and the meridian marking the edge of the map. The angle between grid north and magnetic north is termed the *grid magnetic (GM) angle* (Figure A.15). (You will find this marginal grid symbol on the margin of the map reproduced on the back end paper of your textbook). Thus three kinds of azimuth are also possible on a map; *grid azimuth, true azimuth,* and *topographic magnetic azimuth* (Figure A.15).

Topographic Quadrangles

Most published sets of large-scale topographic maps use the geographic grid to determine the position and size of individual map sheets in a series. A single map sheet, or *quadrangle*, is bounded on the right-hand and left-hand margins by meridians, and on the top and bottom margins by parallels, which are a specified number of minutes or degrees apart.

Seven standard scales comprise the U.S. National Topographic Map Series. English units of length continue to be used, but a change to metric units is in progress.

Series	R.F.	Unit Equivalents
English Units:		
7½-minute	1:24,000	1 in. = 2000 ft
7½-minute	1:31,680	1 in. = ½ mi
15-minute	1:62,500	1 in. = about 1 mi
Alaska	1:63,360	1 in. = 1 mi
30-minute	1:125,000	1 in. = about 2 mi
1-degree	1:250,000	1 in. = about 4 mi
1:1,000,000	1:1,000,000	1 in. = about 16 mi
Metric Units:		
1:25,000	1:25,000	1 cm = 0.25 km
1:50,000	1:50,000	1 cm = 0.50 km
1:100,000	1:100,000	1 cm = 1.00 km
1:1,000,000	1:1,000,000	1 cm = 10 km

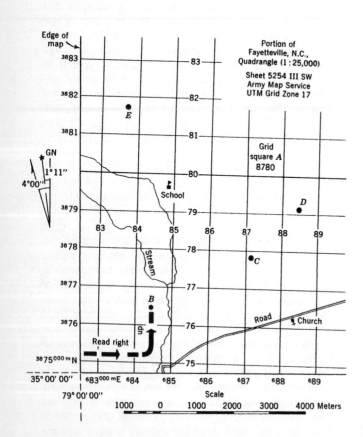

Figure A.14 Grid coordinates on the 1000-meter UTM grid.

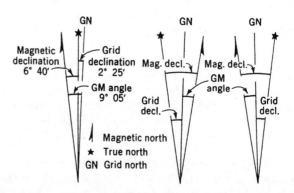

Figure A.15 A special marginal symbol shows the relations among three kinds of north.

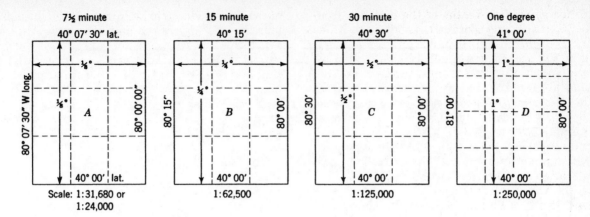

Figure A.16 Large-scale maps of the U.S. Geological Survey are bounded by parallels and meridians to form quadrangles. The four quadrangles shown here represent the scales and areas commonly used in the United States, exclusive of Alaska.

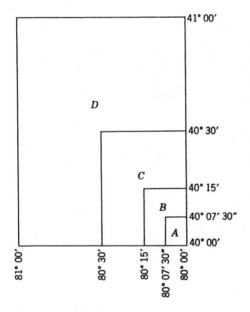

Figure A.17 If the four quadrangles of Figure A.16 were reduced to the same scale, their areas would compare as shown here.

Figures A.16 and A.17 compare the coverages and sizes of standard quandrangles on several of these series.

Topographic Map Symbols

Large-scale topographic maps of the U.S. Geological Survey use a standard set of symbols to show many kinds of features that cannot be represented to true scale. Reproduced on the back end paper of your textbook is the set of symbols used by the U.S. Geological Survey on its current series of large-scale topographic maps, together with a representative example of a map on the scale of 1:24,000.

In general, it is conventional to show relief features in brown, hydrographic (water) features in blue, vegetation in green, and cultural (human-made) features in black or red.

The U.S. Land Office Survey

Topographic maps of the central and western United States show the civil divisions of land according to the U.S. Land Office Survey.

In 1785 Congress authorized a survey of the territory lying north and west of the Ohio River. To avoid the irregular and unsystematic type of land subdivision that had grown up in seaboard states during colonial times, Congress specified that the new lands should be divided into 6-mile squares, now called *congressional townships*, and that the grid of townships should be based on a carefully surveyed east-west base line, designated the "geographer's line." Meridians and parallels laid off at 6-mile intervals from the base line were to form the boundaries of the townships. This general plan, believed to have been proposed by Thomas Jefferson, was subsequently carried out to cover the balance of the central and western states.

The *principal meridians* and *base lines*, from which rows of townships were laid off, are shown in Figure A.18. Principal meridians run north or south, or both, from selected points whose latitude and longitude were originally calculated by astronomical methods. Thirty-two principal meridians have been surveyed. Westward from the Ohio-Pennsylvania boundary, these are numbered from 1 through 6, beyond which they are designated by names.

Through the initial point selected for starting the principal meridian, an east-west base line was run, corresponding to a parallel of latitude through that point. North and south from the base line, horizontal tiers of townships were laid off and numbered accordingly. Vertical rows of town-

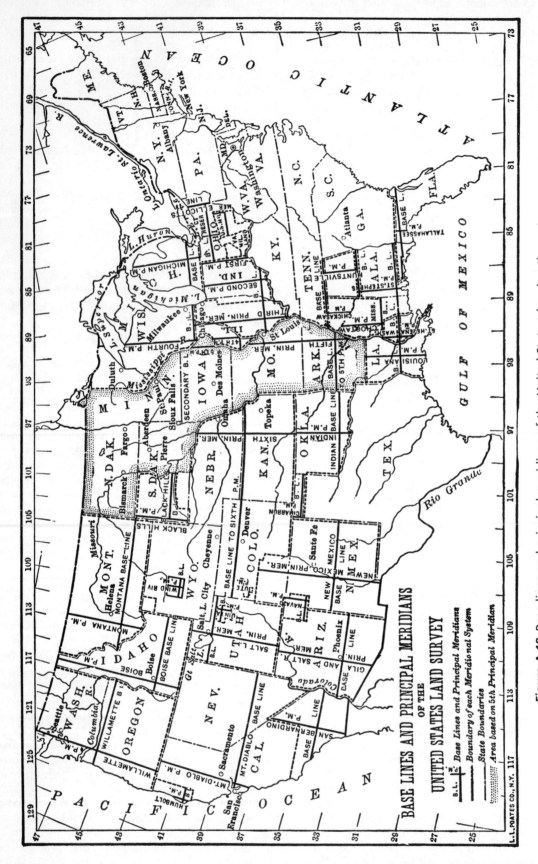

Figure A.18 Base lines and principal meridians of the United States showing the system used by the U.S. Land Office. This map appeared in a popular physical geography textbook by Willis E. Johnson, published in 1907.

368

ships, called *ranges*, were laid off to the right and left of the principal meridians and were numbered accordingly (Figure A.19).

The area governed by one principal meridian and its base line is restricted to a particular section of country, usually about as large as one or two states. Where two systems of townships meet, they do not correspond because each system was built up independently of the others.

Because the range lines, or eastern and western boundaries of townships, are meridians converging slightly as they are extended northward, the width of townships is progressively diminished in a northward direction. To avoid a considerable reduction in township widths in the more northerly tiers, new base lines, known as *standard parallels*, were surveyed for every four tiers of townships. They are designated 1st, 2nd,

3rd standard parallel north, and so on (Figure A.19). The range lines will be shown as offset—to east or west—at the standard parallels. Consequently, roads that follow range lines make a sharp jog when crossing standard parallels.

Subdivisions of the township are square-mile *sections* of which there are 36 to the township. These are usually numbered as illustrated in Figure A.20. Each section may be subdivided into halves, quarters, and half-quarters, or even smaller units. These divisions, together with the number of acres contained in each, are illustrated in Figure A.21.

Suggested reading: *Maps for America: Cartographic Products of the U.S. Geological Survey*, 2nd Ed., by Morris M. Thomson. U.S. Geological Survey, U.S. Department of the Interior, 1981.

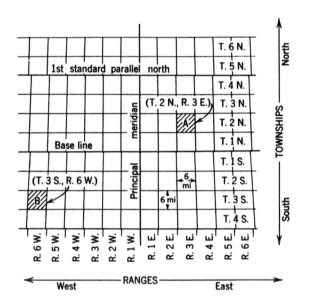

Figure A.19 Designation of townships and ranges.

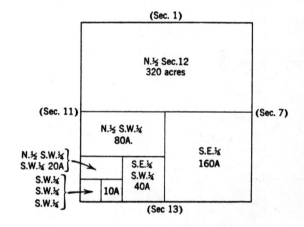

Figure A.21 A section may be subdivided into many units.

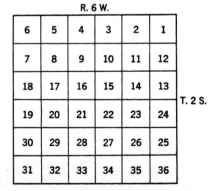

Figure A.20 A township is divided into 36 sections, each one a square mile.